THE ULTIMATE BOOK
OF PUZZLES,
MATHEMATICAL
DIVERSIONS, AND
BRAINTEASERS

Erwin Brecher

THE ULTIMATE BOOK OF PUZZLES, MATHEMATICAL DIVERSIONS, AND BRAINTEASERS

**A Definitive
Collection of the
Best Puzzles
Ever Devised**

St. Martin's Griffin
New York

The following puzzles have been adapted with permission of Sterling Publishing Co., Inc., 387 Park Avenue South, New York, N.Y., 10016 from *Lateral Thinking Puzzles* by Paul Sloane. © 1991 by Paul Sloane.
"The Severed Arm" from "The Arm of the Postal Service"; "The Car Crash" from "The Two Americans"; "A Glass of Water" from "The Man in the Bar"; "The Suicide" from "The Man Who Hanged Himself"; "A Soldier's Dream" from "The Dream"; "The Judgment" from "Another Man in a Bar"; "Death in Squaw Valley" from "Death in Rome"

The following puzzles have been adapted with permission of Sterling Publishing Co., Inc., 387 Park Avenue South, New York, N.Y. from *Challenging Lateral Thinking Puzzles,* by Paul Sloane and Des MacHale. © 1992 by Paul Sloane and Des MacHale. "The Elevator Stopped" from "The Realization"; "The Deadly Scotch" from "The Deadly Party"

Library of Congress Cataloging-in-Publication Data
Brecher, Erwin.
The ultimate book of puzzles, mathematical diversions, and brainteasers / by Erwin Brecher.
p.   cm.
ISBN 0-312-14143-2
1. Mathematical recreations.   I. Title.
QA95.B74   1996
793.73—dc20   95-42895
CIP

First published under the title *Journey Through Puzzleland* in Great Britain by Pan Books

First U.S. Edition: May 1996

10 9 8 7 6 5 4 3 2 1

To the memory of

FRED AND HARRY

**puzzle-solvers supreme,
with whom I spent many
stimulating hours.**

Contents

Acknowledgments

Little can be achieved in any field of human endeavor without the support, encouragement and assistance of others. *The Ultimate Book of Puzzles* has been no exception; it seemed a most formidable task when I first put pen to paper.

The puzzles have different origins; many are of my own construction, while the "Golden Oldies" have been carefully selected to offer variety and curiosity, and to challenge the puzzle-solving capabilities of the most accomplished devotee. I am indebted to many copyright holders for permission to use material from their books. Sources include: Scot Morris, whose *Omni Games* proved to be a rich source; Martin Gardner, the doyen of puzzlists; Victor Serebriakoff and Octopus Publishing; Emerson Books Inc.; Philip Carter, Ken Russell and Sphere Books; Wide World/Tetra; *Encyclopedia Britannica; The Little, Brown Book of Anecdotes;* Sterling Publishing Co., Inc., 387 Park Avenue South, New York, NY 10016, for material from *The World's Best Puzzles* by Charles Barry Townsend, © 1986 Charles Barry Townsend; *The Moscow Puzzles* by Boris A. Kordemsky, edited by Martin Gardner, translated by Albert Parry (Penguin Books, 1975), © Charles Scribner's Sons, 1972; *Riddles in Mathematics* by Eugene P. Northrop; *Classic Puzzles* by Philip Carter and Ken Russell, published by Blandford Press; Christopher Maslanka, for material from *The Guardian Book of Puzzles* (Fourth Estate, 1990); Michael Holt, for material from *More Math Puzzles and Games;* Paul Sloane, for material from *Lateral Thinking Puzzles;* Pierre Berloquin for material from *Games of Logic* (HarperCollins, 1980) and Gyles Brandreth, for material from *Classic Puzzles and Word Games.* All these books are highly recommended and are a must for any puzzle library.

My thanks go to Mike Gerrard for verifying the mathematics and physics aspects of some of my puzzles; to my son Michael, who edited the manuscript and made many valuable suggestions; to Les Smith, who rescued me from drowning in paperwork; and to Jennifer Iles, who coped with innumerable corrections and revisions.

Foreword

Erwin Brecher's book is a fascinating collection of old and new brain teasers covering the whole spectrum of this challenging field but avoiding repetitions and trick questions so often found in puzzle literature.

A particularly attractive feature is the author's approach to presenting solutions. Not only does he describe a novel technique to solve a specific type of geometrical problem, which he calls the Zero Option, but he also guides the reader through the thought process leading to the solution. This is a welcome change to many puzzle books which simply state the answer, leaving the reader baffled as to how the solution was arrived at and more than a little frustrated at being unable to connect. The chapter on the Fourth Dimension is a thought-provoking effort at proving, in easy language, that this mysterious concept represents something real in our universe.

The Ultimate Book of Puzzles is a valuable addition to any puzzle library.

Victor Serebriakoff

Hon. President of International Mensa

Introduction

This book is not intended for the serious mathematician. We will not be discussing diophantine equations, fibonacci numbers or the Fermat theorems. Rather, it is written and compiled as entertainment for people who enjoy putting their mental agility to the test, a hobby gaining in popularity to judge from the growing number of periodicals that carry regular puzzle features alongside their chess, crossword and bridge columns.

Over sixty years of puzzle-solving, I have developed a prejudice in favor of puzzles presented as simply as possible and against unnecessary embellishment. I have imposed that prejudice on my readers. In these pages, you will not be meeting too many kings, queens or knaves, or journeying to mythical lands in search of dragons. Clear explanations for all the puzzles, including those in this introduction, begin on page 131.

I was first bitten by the puzzle bug in the mid-1920s, in Vienna. At the close of a geometry lesson, our professor gave us the following problem to solve as homework:

i. Area of a Circle

"You know the formula for the circumference of a circle to be $2\pi r$," the professor said. "Find the area, using only the most elementary geometrical formulae."

At that age, we had only just heard about calculus, that mysterious product of mathematical genius. I spent the best part of the afternoon trying to find the answer. Suddenly it struck me. The solution was simple and yet so elegant that I felt elated for the rest of the day.

I have been an addict of puzzles, mathematical diversions and brainteasers ever since.

Over the years, I have been an avid reader of puzzle literature: writers such as H. E. Dudeney, Scot Morris, Martin Gardner, Sam Lloyd, James Fixx; publications like *Scientific American* and *Omni*. At the same time, I recognize that textbook mathematics can be very boring, and lectures on the subject are often prescribed as a cure for insomnia! There is the anecdote of the professor who was droning away on the theory of probability, while his students became progressively more restive or began yawning and dozing off. After the lecture, a colleague asked him whether he found it upsetting when his students started checking their watches demonstratively. "Oh, no," the professor replied. "Only when they start shaking them."

Puzzle fans will know the feeling of exhilaration which accompanies the solving of a problem, particularly if the solution is elegant. I would not go quite as far as Sigmund Freud, who believed that solving a puzzle provided a thrill mildly akin to sexual gratification, but it does, at least, produce what Freud termed *Lustgefuehl,* a feeling of intense pleasure.

Thousands of books have been published on the subject of puzzles and mathematical diversions, by the authors mentioned above and others. While many of the puzzles in this volume have been devised by me, I have also borrowed freely from other sources. By way of an apology, I would point out that the totally original puzzle is a thing of great rarity. Many of the best puzzles have been handed down from generation to generation, modified somewhat, perhaps; progressively refined. But their origins are no longer ascertainable and they are not what lawyers would describe as copyrightable. By way of general credit, let me just say that I have derived many hours of pleasure from the works of all the above-named authors and others.

In an article called "The Psychology of Puzzle Crazes" (*Nineteenth Century Magazine,* December 1926), the great British puzzlist Henry Ernest Dudeney made two complaints. "The literature of recreational mathematics," he said, "is enormously repetitious, and the lack of an adequate bibliography forces enthusiasts to waste time in devising problems that have been devised before." This is no longer the case, thanks to an exhaustive bibliography published in several volumes by the National Council of Teachers of Mathematics (USA); the bibliography on page 263 is a small selection from this.

My purpose has been to assemble a distillate of the finest puzzles, mainly of the "one-off" type, that I have come across over more than half a century. Their

Introduction

solution requires at most only an elementary knowledge of mathematics (specifically geometry), but, usually, a high degree of inspirational intelligence.

To this end, I have limited or excluded altogether many types of brainteaser commonly found in puzzle books—magic squares and anagrams, for instance; decanting problems and "crossings" puzzles, palindromes and mazes—where finding the solution is purely mechanical or a pedestrian, trial-and-error exercise. However, some apparent trial-and-error puzzles are nothing of the kind:

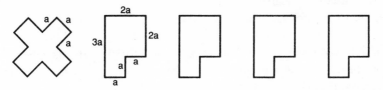

Fit the five shapes illustrated above together to form a perfect square with no gaps.

Of course, one could cut out the shapes and play around with them until a little judgment and a lot of luck yields the square. Or one could use logic. The cross (first shape) can be divided into 5 small imaginary squares, as—in a different way—can the other four shapes. This means the final square will consist of 25 of these small squares; that is, each side will be 5 little squares long. Knowing that, one can figure out how to rearrange the shapes to obtain straight edges of the required length.

I have also avoided trick questions. Many are trivial. Two of the worst examples:

A train travels from London to Liverpool at 80 miles per hour. At the same time, another train leaves Liverpool for London, traveling at 95 miles per hour. Which train will be nearer Liverpool when they meet?

and:

To test your knowledge of international law: if an airplane crashes on the border between the USA and Canada, in which country would the survivors be buried?

Others are borderline.

ii. Steamship

A steamship leaves Hamburg, sailing due north. Two seamen, Jim and John, are on deck. Jim is facing east and John is facing west. Suddenly, without turning around, John says: "Jim, your nose is bleeding." How could he tell?

Puzzles concerning liars and truthtellers are legion, and have become pretty tiresome. They enjoy differing degrees of complexity, but solving them gives little intellectual satisfaction.

   In their purest form, one might go like this:

You meet two people, one belonging to a tribe of truthtellers, the other to a tribe of liars. Formulate one question, addressed to either, which will tell you which is which.

Or, a little harder:

You meet two people, a liar and a truthteller, without knowing which is which. Formulate a question, to be asked of only one of them, to which the answer must be "Yes" regardless of whether you ask the liar or the truthteller.

Harder still (from *Alice and the Forest of Forgetfulness*):

iii. The Lion and the Unicorn

The Lion and the Unicorn were frequent visitors to the forest. These two are strange creatures. The Lion lies on Mondays, Tuesdays and Wednesdays, and tells the truth on the other days of the week. The Unicorn, on the other hand, lies on Thursdays, Fridays and Saturdays, but tells the truth on the other days of the week. One day Alice met the Lion and the Unicorn resting under a tree. They made the following statements:
LION: Yesterday was one of my lying days.
UNICORN: Yesterday was one of my lying days too.
   From these two statements, Alice was able to deduce the day of the week. What day was it?

Often, solving puzzles reduces to a tedious reliance on algebraic equations. For instance:

"How much do you pay for these cucumbers?" asked the inquisitive visitor.

"Well," was the reply, "I pay just as many drachmas for six dozen as I get cucumbers for 32 drachmas."

What was the price per cucumber?

Almost as often, the puzzle's setter gives no hint of the equation necessary to solve it—the published answer to the above puzzle (adapted from *Puzzles and Curious Problems* by H. E. Dudeney), for example, is $\frac{2}{3}$ drachma. This is absurd, since such interest as there is in these puzzles lies wholly in constructing the correct equation to yield a solution; working the equation out is a purely mechanical exercise.

The following is one of the better exemplars of this type of puzzle:

iv. The Two Friends

Two friends meet to celebrate their birthdays, which fall on the same date.

"This is a very special birthday," remarks one of them, "because, while we are together 63 years old, I am also twice as old as you were, when I was as old as you are now."

Find their ages.

Paradoxes are not puzzles in the true sense but are often thought-provoking:

In a town that boasts only one barber, all the men fall into one of two groups: those men who are shaved by the barber and those who shave themselves. To which group does the barber belong?

and:

A Cretan by birth, Epimenides maintained that all Cretans are liars, a statement that, if true, makes the speaker a liar for telling the truth.

THE ZERO OPTION

Before continuing with a more complex paradox I have to explain a technique I have developed which can be applied to certain problems, whose solution conventionally involves a laborious process but which, using the Zero Option, can be solved instantly.

It is not always easy to recognize problems which will lend themselves to this technique. To some extent, recognition is intuitional. One telltale sign is an apparent lack of sufficient data, which implies that the solution is independent of the missing information. On this assumption the relevant dimension can be taken as zero. This transforms the puzzle from an arduous exercise in geometry into what Charles W. Trigg, Dean Emeritus of L.A. City College, calls a "quickie," a term he coined to describe problems that yield almost instantly to a flash of inspiration.

To my knowledge, the Zero Option has not been used before, or at least not with any appreciation of its significance.

v. Galileo's Paradox

Take any square ABCD and draw the diagonal BD.

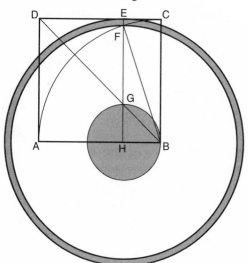

With B as the center and with radius BC, describe the quarter-circle CFA. Draw any line HE parallel to BC, intersecting the quarter-circle at F and the diagonal at G. With H as centre, construct circles with radii HG, HF and HE respectively.

It is not difficult to show that the area of the shaded circle is equal to that of the shaded ring. To do so, note first that triangle FBH is a right-angled triangle. Consequently, by the well-known Pythagorean theorem:

$$BF^2 = HB^2 + HF^2$$
$$or \quad HB^2 = BF^2 - HF^2$$

Now, HE = BC, and BC = BF (both radii of the same quarter-circle). Thus, HE = BF. Similarly, HB = HG (radii of the same circle). So, the above equation could be written as:

$$HG^2 = HE^2 - HF^2$$

Multiplying both sides of the equation by π:

$$\pi.HG^2 = \pi.HE^2 - \pi.HF^2$$

The left-hand side of this equation represents the area of the shaded circle. The right-hand side, being the difference of the areas of the circle with radii HE and HF, represents the area of the shaded ring.

Using the Zero Option principle, let HE move to the right and approach the position BC. As HE coincides with BC, the shaded circle shrinks to the point B and the shaded ring shrinks to the circumference of a circle with HE (now BC) as radius. But since the areas of the shaded circle and the shaded ring are equal for any position of HE, we must conclude that a single point is equal to the circumference of a circle.

The following puzzle meets the Zero Option criterion, inasmuch as the diameter of the hole is not given.

vi. Hole in the Sphere

A cylindrical hole, 6 inches long, has been drilled through the centre of a solid sphere. What is the volume remaining in the sphere?

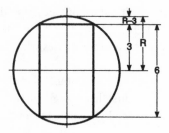

(The solution is given on page 135 both in its standard form and using the Zero Option.)

Here is another example:

vii. Rope around the Equator

A thin rope is stretched around the equator, at a distance of 1 foot (suitably supported). The supports are then collapsed, and the rope is tightly drawn around the equator. Before looking at the answer, try to estimate the length of rope you would be left with.

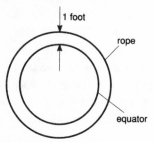

(The circumference or radius of the equator is not given, and most readers will not happen to have this information to hand. It would certainly seem that this puzzle is a case of "insufficient data.") The Zero Option is in some respects similar to the "method of exhaustion" originated by Eudoxus of Cnidus (400–437 BC), and extensively applied by Archimedes. This method was used not only to determine areas and volumes of geometrical figures, but also to calculate π, which is the constant ratio between the circumference of a circle and its diameter.

This was done by a succession of inscribed and circumscribed polygons, starting with squares and then doubling the sides to get an octagon. The figures below demonstrate the principle involved.

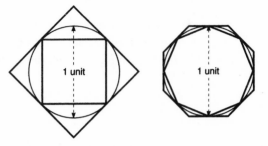

This method of doubling sides can continue, until the perimeters of the two polygons approach ever closer the circumference of the circle.

A similar approach can be used with the Area of a Circle problem on page xiii.

I know of no better way to demonstrate the beauty of an ingenious answer to a neat puzzle—and a Zero Option attitude to problem-solving—than to end by retelling a famous anecdote concerning the mathematician Carl Friedrich Gauss.

Born in 1777 in Brunswick, Gauss showed early and unmistakable signs of being an extraordinary youth. As a child of three, he was checking—and occasionally correcting—the books of his father's business; this from a lad who could barely peer over the desk top into the ledger!

viii. Gauss's Problem

One day, one of Gauss's elementary school teachers asked the class to work quietly at their desks and add up the first 100 whole numbers. That'll keep the little devils busy for a while, the teacher assumed. But hardly had the teacher finished speaking—and hardly had the other children progressed beyond 1 + 2 + 3 = 6—when Gauss walked up to the teacher's desk and handed him the answer. The teacher didn't know whether to be more annoyed by Gauss's impertinence, or by the fact that his answer was correct.

(For the record: it is true that people are occasionally born with the phenomenal ability to perform mathematical functions in their heads at lightning speed; but Gauss was not one of them.)

Can you do as well?
In 10 seconds?!

Fallacies are similar in nature to paradoxes, and refer to statements based on mathematical, geometrical or pseudoscientific conclusions which are absurd and clearly flawed, though the source of error is not always readily identifiable.

The flaw in the thought-process in the following conundrum concerning the date-line is quite obvious, though nonetheless amusing:

The date-line is "a north-south line through the Pacific Ocean where, by common usage, the date changes. East of the line it is one day earlier than it is to the west. The line is necessary because the Earth is divided into 24 one-hour time zones (of 15° longitude each) which make one full day on the Earth. Since the Earth rotates toward the east, a given clock time progresses westward around the world. Thus noon arrives at London (0° longitude) 5 hours before it does in Washington (75° west of London), 8 hours before it does in San Francisco (120° west of London), and 12 hours before it does at 180° west of London. At that longitude, it would be midnight when it was noon in London. However, if longitude and time were counted east from London's noon, though midnight would be approached again, time would be later in the day, one hour for each 15°, rather than earlier. This situation would place the same day on both sides of 180°. This cannot be, for the new day that starts westward from the 180th meridian must have a date one day later than the day which has just arrived there from the east. The anomaly is solved by arbitrarily designating where a new day begins. On both sides of the line the time of day is the same, but the name of the day, and hence its date, is changed—forward if going west, backward if going east."

In other words, a traveler crossing the line to the west would lose a day, and going east, gain a day. Using a spacecraft capable of circling the Earth in, say, 6 hours, one could, depending on the direction, go back into history or forward into the future, by gaining or losing 18 hours with every orbit.

Here are three more challenging examples of apparent fallacies:

ix. Let A = B + C

This is designed to show that two unequal numbers are nevertheless equal (no, the trick is not that B or C = 0!).

Multiply both sides by $(A - B)$ to obtain:

$$A^2 - AB = AB + AC - B^2 - BC$$

Moving AC to the left side results in:

$$A^2 - AB - AC = AB - B^2 - BC$$

Factor:

$$A(A - B - C) = B(A - B - C)$$

Divide each side by $(A - B - C)$ to get:

$$A = B$$

Find the source of error.

x. Casino Game

In *Riddles in Mathematics,* Eugene P. Northrop contemplates this sure-fire system for success at the roulette table (ignoring zero):

Stick to one table and bet on a series of consecutive spins of the wheel. Begin by staking 1 dollar on any of the 50–50 bets (black-red, odd-even, first 18–last 18). If you win you are ahead 1 dollar, and that series of plays is over. If you lose, stick to the same bet, staking 2 dollars. If you win this time, you are 1 dollar ahead. So long as you lose, keep to the same bet, doubling your stake each time until you win, at which point you will be 1 dollar ahead.

Whenever you win, you will be 1 dollar ahead; then begin the series again (at 1 dollar) until you win a second dollar; and so on.

All very well, you say. But I'm not a billionaire; the bank could easily wipe me out before I win. In theory, yes, but—as Northrop points out—in practice not so. Successive runs of more than 10 to 12 are extremely rare, which means that you will likely not need more than 4–5,000 dollars to make Northrop's system work for you.

Now consider this: Northrop's view of the rarity of long runs is borne out by the statistics which casinos publish themselves. Let us assume you visit a casino and wait until the same color has come up at a particular table, say, 15 times in a row. The published statistics tell you that a run of 15 happens only once in about 6 years at the casino. You tell yourself this means that the chances overwhelmingly favor the *other* color coming up on the next play. Are you right?

xi. Jack and John

Two friends, Jack and John, argue as to who owns the more expensive tie. They finally agree to settle the argument by checking with the store where both ties were bought. To make it more exciting, they agree on a bet, whereby the man who wins—i.e. who owns the more expensive tie—will give it to the loser, by way of consolation.

Both are pleased with the arrangement, reasoning as follows: "If I win, I will be poorer by the tie I am wearing, but if I lose, I will win a more expensive tie. As the chances are obviously equal, the bet is clearly to my advantage."

How can a bet be favorable to both parties?

I include only two alphametic items; some readers will enjoy these code-breaking puzzles, though the solving technique is mechanical and, once mastered, repetitious. One example, The Long Division, can be found on page 43; the other, rather neat, example is as follows.

xii. The Fax

A fax received by John's parents reads:

$$\begin{array}{r} S\ E\ N\ D \\ M\ O\ R\ E \\ \hline M\ O\ N\ E\ Y \end{array}$$

Each letter stands for a digit, and for the same digit throughout the text. Different letters stand, of course, for different digits. The result will show that the two numbers replacing the words SEND and MORE are correctly totaled in the number which replaces MONEY.

One of the basic criteria of any good puzzle is that the question or answer not be deceptive. When they are, the result is not quite a trick question, but certainly a letdown:

Six pails stand in a row. The first three are full of water, the second three are empty. Moving only one pail, line them up so that the full and empty pails alternate.

The phrasing of the third sentence above proves to have been misleading (and the puzzle a rather disappointing "fraud") when one discovers that the solution is to pick up the second pail on the left and pour its water into pail No. 5. The next one is a little more intriguing:

xiii. Where There's a Will

A Middle Eastern potentate died, leaving 17 camels. His will specified that they be divided among his 3 sons as follows:

$\frac{1}{2}$ to the oldest son;

$\frac{1}{3}$ to the second son;

$\frac{1}{9}$ to the youngest son.

The sons were puzzling over how this could be done when a wise man happened to ride by on a camel. How did he solve their problem?

Series belong in the category of trial-and-error problems. They are often based on differentials or products, made only slightly more difficult by adding and/or deducting a constant.

I have found few examples which provide any kind of intellectual satisfaction, though the following are, perhaps, borderline:

xiv. Number Series

How are the following numbers arranged?

$$0, 2, 3, 6, 7, 1, 9, 4, 5, 8$$

xv. Next

What are the next four numbers in this series?

$$12, 1, 1, 1, 2, 1, 3, ????$$

Match or toothpick problems are usually matters of trial and error; only a couple have found their way into *The Ultimate Book of Puzzles*. In addition, this one is good fun:

xvi. Move One Match

Move just one match to restore the equation:

$$VI = II$$

Identification problems have become extremely popular. Many readers, it seems, can happily spend hour after hour figuring out "who does what," "who sits next to whom," or "who is the murderer," even though these are probably the least imaginative items in puzzle literature. None are included in the main text of *The Ultimate Book of Puzzles,* though the following, from *101 Puzzles in Thought and Logic* by C. R. Wylie Jr., is one of the neatest I have seen:

xvii. Artistic Fields

Boronoff, Pavlow, Revitsky and Sukarek are four talented creative artists—one a dancer, one a painter, one a singer and one a writer (though not necessarily respectively).

1. Boronoff and Revitsky were in the audience the night the singer made his debut on the concert stage.
2. Both Pavlow and the writer have sat for portraits by the painter.
3. The writer, whose biography of Sukarek was a best-seller, is planning to write a biography of Boronoff.
4. Boronoff has never heard of Revitsky. Can you work out who is who?

The subject of probability is at quite the other end of the spectrum. The reason I have included only a few, rudimentary probability problems in this volume is because the topic, once embarked upon, can quickly take over an entire book. Indeed, several excellent, in-depth studies have been published. As a taster to this most fascinating area of recreational mathematics, I offer the following:

xviii. The Joker

A friend offers you the following wager. He takes three playing cards from a new pack—one joker and two aces—then deals them out face down in front of you. He then asks you to remove one card from the group of three, keeping it face down so that neither of you can see it.

The object of the game is for you to try to pick the joker. If you succeed, he will pay you £1; if you fail, you lose £1. The card you picked was at random because you didn't know in which order your friend dealt them out. But he did; he knows exactly which card is which, and turns over one of the two cards left on the table, showing it to be an ace.

"Now there are two cards left," your friend says. "One on the table and one in your hand. One is another ace; the other is the joker."

What are the odds of the card in your hand being the joker?

Before concluding I want to remind you that solutions to all puzzles in this introduction, prefixed with a Roman number, are given in the section starting on page 131.

Erwin Brecher

PUZZLES

1. Puzzles in Geometry

"Let no one enter who does not know geometry."

—Inscription on Plato's door

1. Tangent Triangle

In the figure below, two tangents to the circle have been drawn from point C. The lines YC and XC are equal, each having a length of 10 units. Line AB was then drawn, tangent to the circle at point P (point P being a point on the circumference of the circle randomly chosen between points X and Y). Can you calculate the length of the perimeter of the triangle ABC?

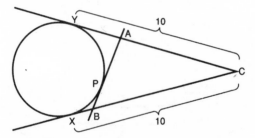

Because, at first glance, there seems to be insufficient information given, this puzzle is a natural for the Zero Option approach.

2. Area of Overlap

The figure below shows two squares. The sides of the smaller square are 3 inches in length, those of the larger square, 4 inches. Point D is the corner of the large square and the center of the small square. The large square is rotated around point D until the length of AB is 1 inch and BC 2 inches. Can you calculate the area of BCED, the overlap of the two squares?

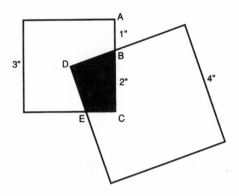

3. The Pond

In a garden, there is an elliptical pond whose area is exactly 50 square feet. Along the border of the pond is a flower bed 2 feet wide, and bordering the flower bed is a path also 2 feet wide (see figure below). The entire region may be assumed to lie in a plane.

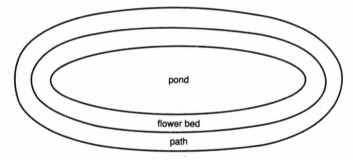

Can you tell the area enclosed by the largest ellipse?

4. Two Angles

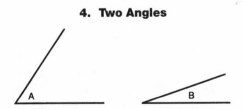

I have just constructed these two angles (see overleaf), and I can prove that angle B is exactly one-third of angle A. Yet I performed the construction in a finite number of steps using only an unmarked straight-edge and a compass. Furthermore, angle A does not have some special value, such as 45°, which would make the construction possible. On the other hand, it is generally accepted that it is impossible to trisect an angle under the conditions stated.

How was the construction performed?

5. Area of Annulus

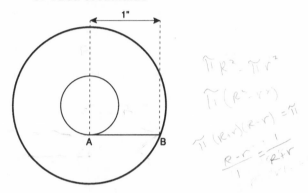

In the figure above, line AB, which is 1 inch long, is tangent to the inner of two concentric circles at A and intersects the outer circle at B. What is the area of the annular region between the circles?

Try the Zero Option and compare with the standard solution.

6. The Round Window

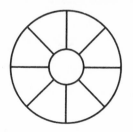

A designer is planning a 9-paned round window for a church on the plan shown above. He decides that the window would be aesthetically correct if the area of each of the 8 outer panes were equal to the area of the circular inner pane. Assuming the inner circular pane is 2 feet in diameter and the thickness of the wood between the panes can be ignored, what should be the length in inches of the spokes which separate the outer panes?

7. The Hypotenuse

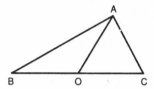

A line is drawn from the corner of a right-angled triangle (above) to the middle of the hypotenuse. Can you prove that its length is half the hypotenuse?

8. Hex Sign

The hex sign above is made up entirely of circles 1 inch in radius. Can you calculate the area of each of the shaded petal-shaped areas?

9. How Large is the Cube?

What size cube has a surface area equal (in number) to its volume?

10. As the Fish Swims

At the center of a Municipal Park is a large, circular pool. The Civic Club decides to stock the pool with fish. One goldfish is placed into the water at the
edge of the pool. It swims due north for 60 feet before running into the pool's
edge. It then swims east and hits the edge again after swimming 80 feet. What
is the diameter of the pool?

11. Pentagram

The pentagram above has five points. Can you work out the angle at any one of
the five points of a regular pentagram?

12. Diagonal Problem

In the diagram below, a rectangle ABCD has been inscribed in the quadrant of
a circle.

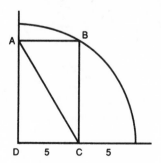

Given the unit distances indicated, can you determine the length of the diagonal AC?

Try also the Zero Option.

13. Squares from a Square

Using only a straight-edge, construct 5 smaller squares, the areas of which will total that of the larger square.

14. The Goat

A villager pays rent to a local farmer who allows him to graze his pet goat on his (the farmer's) field. When on the field, the goat is tethered to a post by a 21-foot rope, so that the area on which it grazes is a circle with a radius of 21 feet. The farmer has decided to build a shed on the field, 14 feet by 7 feet, using the post as one of the corner posts of the shed (see below).

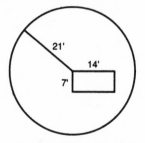

If the farmer has been charging the goat's owner $100 per annum, by how much should the rent be reduced to reflect the grazing area lost?

15. Water Lily

A water lily stands 10 inches above the surface of the water (see below). If it was pulled over until the head touched the surface it would disappear at a point 21 inches from where it was originally.

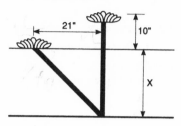

How deep is the lake at that point?

16. The Sphere

A sphere with a radius of 1 meter has been rolled into the corner of a room so that it is tangent to two walls and the floor (see below).

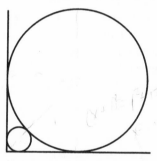

What would be the radius of a smaller sphere that is tangent to the same two walls and floor and touches the large sphere as well?

17. Division of Land

A man leaves the piece of land (see overleaf) to his 4 sons, with instructions that it should be divided up into 4 equal pieces, each having the same shape as the original piece of land. How can this be achieved?

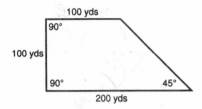

18. Center of a Circle

Find the center of the circle using only the draftsman's triangle and pencil, as shown above.

19. The Window

A modern country house has a gable equilateral-triangular in shape in which the owner wants to cut a square window as big as possible (see below).

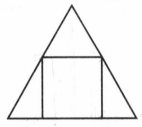

Ignoring frames, etc., what is the area of the window compared to the gable?

20. The Arch

The figure below shows a typical Gothic arch.

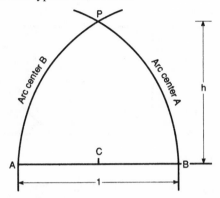

What is *h* equal to? Try to solve it in your head in 2 minutes without drawing anything.

21. Equal Areas

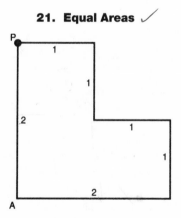

Divide the rectilinear right-angled figure shown above into two equal areas with one straight line PP_1.

22. Angular Problem

Without using trigonometry, prove that angle C in the figure below equals the sum of angles A and B.

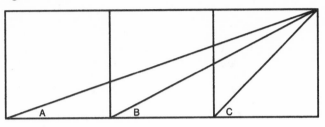

23. Overlapping Circles

In the figure below, 3 equal circles have been drawn so that each one passes through the centers of the other two. Is the area of overlap, shown shaded in the diagram, more or less than a quarter of the area of a circle?

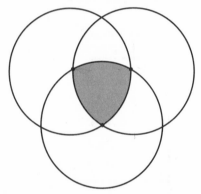

This problem can be solved without the need to find the area of the equilateral triangle inscribed in the overlap and then adding the areas of the 3 segments of circles on each side of the triangle. In fact, no geometrical formulae are required at all.

24. Sideways

The figure on the overleaf shows a monument viewed from the front and from above. What does it look like from the side?

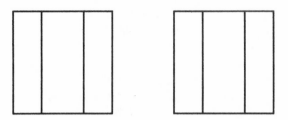

25. Three-dimensional Object

Shown below are two different views of the same three-dimensional object—one from the front and one from the side.

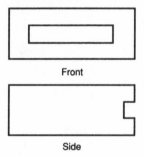

What is the simplest three-dimensional shape that would produce these views?

26. Circles in the Squares

Opposite is a succession of squares within circles, each just touching the other, so that you have, in the order from outside in, circle-square-circle-square-circle.

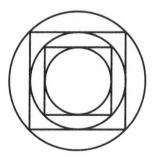

If the diameter of the outside circle is 10 inches, what is the diameter of the inside circle? (You can solve this in your head.)

27. Spider and Fly

The geodesic problem of the spider and the fly was probably Dudeney's most famous puzzle. It was first published in 1903, and aroused widespread public interest in 1905 when it appeared in the London *Daily Mail.*

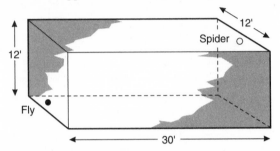

A rectangular room is 30 feet long, 12 feet wide and 12 feet high, as shown above. The spider is sitting 1 foot down from the ceiling at the center of one of the end walls. The fly is 1 foot above the floor at the center of the opposite end wall, and being petrified with fear is unable to move. What is the length of the shortest route the spider can take to get to the fly?

28. Overlapping Areas

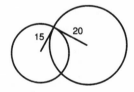

A circle of radius 15 units intersects another circle, radius 20 units, at right angles (see above). What is the difference of the areas of the non-overlapping portions?

29. The Vanishing Square

Design a square as shown in figure A, with an area of 64 units. Cut the board along the lines indicated and rearrange as shown in figure B, which yields a rectangle with an area of 65 units.

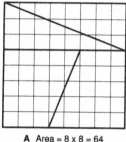 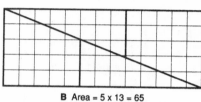

A Area = 8 x 8 = 64   B Area = 5 x 13 = 65

The gain of one square can be explained by a long, thin, diamond-shaped gap along the diagonal of the 5×13 rectangle. This is hardly noticeable to the naked eye.

The story goes that a similar optical illusion was used by clever forgers to make 15 $20 bills out of 14 bills. How was this done?

30. The Hula Hoop

Consider a girl whose waist is exactly circular, not smooth, and temporarily at rest. Around her waist rotates a hula hoop of twice the waist's diameter. Show that after one revolution of the hoop, the point originally in contact with the girl has traveled a distance equal to the perimeter of a square circumscribing the girl's waist.

31. Bisect a Line

Assume you want to bisect a straight line, without any tools except a pair of compasses. How would you do it?

32. Centers of Three Circles

If they are connected, the centers of any three circles that touch one another will form a triangle, regardless of the sizes of the circles. An example of this is given on the overleaf.

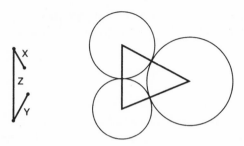

1. Can such triangles be made of any proportion?
2. When the sides X, Y and Z of a figure are specified, if Z is greater than X + Y, a triangle cannot be made because X and Y cannot meet regardless of the angles.

    Can this be true in the case of a triangle formed by three circles, and if not, why not?

33. The Magnifying Glass

I drew a figure 8 on a piece of paper and then looked at it through a magnifying glass: it looked twice as big. I then drew another figure and looked at it through the magnifying glass: it looked no bigger than before. What had I drawn?

34. Isosceles Triangle

Each of the two sides of an isosceles triangle is 10 inches long. Find the length of the third side that maximizes the area of the triangle.

35. Two Hexagons

Regular hexagons are inscribed in and circumscribed outside a circle, as shown above. If the smaller hexagon has an area of 3 square inches, what is the area of the larger hexagon?

36. The Flag

A flag is 4 feet wide and has a design with a red square on a white background, though the white only shows at the sides (see the figure below).

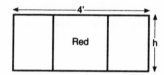

How big should the red square be to give the greatest amount of white on the sides? The width of the flag has to be 4 feet, of course. Without calculus, what value of h gives the maximum of white?

37. Tangent Circles

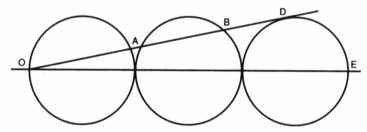

In the figure above, three tangent circles of equal radius r are drawn, all centers being on the line OE. From O, the outer intersection of this axis with the left-hand circle, line OD is drawn tangent to the right-hand circle. What is the length, in terms of r, of AB, the segment of this tangent which forms a chord in the middle circle?

38. The Avenue

The local residents' association has been complaining about the plans for a new housing estate in the town. The proposed site, as shown on the overleaf, is a perfect square, each side being $\frac{3}{4}$ of a mile.

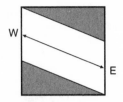

Houses are to be built only on the shaded areas, that is, the two identical triangular areas. The space in between, down the center of which an avenue runs from west to east, is to be occupied by a communal garden.

It is the devotion of so much space to this garden—it occupies $\frac{7}{12}$ of the total area of the estate—that has provoked adverse criticism. What is the length of the central avenue?

39. Four-wheel Cart

A horse-driven 4-wheel cart drives round in a circle. The wheels are 5 feet apart on the axles, and the outer wheels are making 2 turns to any single turn made by the inner wheels. Assume the circle is as tight as possible, without slip of the outer wheels. What is the circumference of the circle described by the outer wheels?

40. Crossing the River

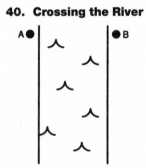

During the war, a dispatch rider arrived at point A on a river which he had intended to cross (see above). However, he found that the bridge had been destroyed and all that was left were the steel girders at A and B. To work out a means of crossing the river he had to calculate how wide it was. How did he do this?

41. Wheels and Spheres

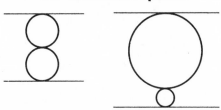

The left-hand figure above shows the side view of two equal-sized wheels between two parallel rails. Assuming no slippage, the wheels can roll either way and still maintain their relative positions vertically. The same would be true if the wheels were spheres and the rails were planes.

However, if the top sphere were much larger than the sphere below, as in the right-hand figure above, and they were pushed to one side, which sphere would go ahead of which?

42. The Shed

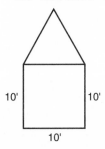

When a cubical shed with a pyramidal roof is viewed from a point opposite the center of a side, it appears as a square with an equilateral triangle on top, as shown in the figure above. Given that the shed is 10 feet wide, calculate the area of the roof.

43. Two Gold Coins

A man had a desk with an exactly circular hole which had once held an inkwell. He had two gold coins of the same thickness. The first coin exactly fitted the hole. The smaller coin, when slid gradually over the hole, tipped into it when its edge reached the center of the hole. If the larger coin weighed 6 ounces, what was the weight of the smaller coin?

44. Pieces of Paper

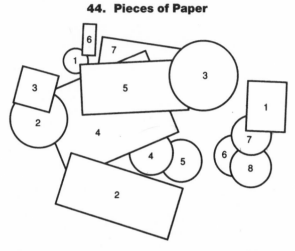

The figure above shows 7 paper rectangles and 8 paper circles lying on a flat surface. Each corner of a rectangle and each point where edges intersect is a point.

This configuration contains 6 sets of 4 cocyclic points; that is, points that can be shown to lie on the circumference of a circle. Since the corners of any rectangle lie on a circle, 4 of the sets are immediately identifiable as the corners of the 4 rectangles not covered by other shapes. Can you find the other 2 sets?

45. Balloon and Box

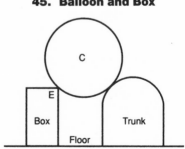

The figure on the overleaf shows a balloon resting on a box at the left and a trunk with a hemicylindrical lid at the right. The balloon, whose outline may be considered a circle, leaks and shrinks. As it sinks and falls, what is the locus of its center, point C? (The edge of the box, E, may be considered a point.)

46. Find the Area

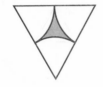

The tessellated figure above consists of 3 equal arcs in an equilateral triangle. Each side measures 2 feet. What is the area of the shaded part?

47. Self-congruent

A straight line is called *self-congruent* because any portion of the line can be fitted exactly to any other portion of the same length. The same is true of the circumference of a circle, because any part of the circumference is exactly like any other part of the same length. By contrast, an oval line is not self-congruent, because parts of it have different curvature; a portion taken from the side would not fit a portion at one of the ends.

Can you think of a third type of line that is self-congruent?

48. Two to One

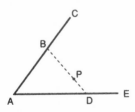

Given an acute angle CAE and a point P within the angle (above), use a compass and straight-edge to construct a segment BD passing through P, such that BP and PD stand in the ratio 2 : 1, B and D lying on CA and AE respectively.

49. Trisecting the Square

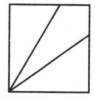

From one corner of a square (above), extend 2 lines that exactly trisect the area of the square. In what ratios do these lines cut the 2 sides of the square?

50. The Circular Table

A large circular table is pushed into the corner of a room, so that it touches both walls. On the circumference is a spot exactly 8 inches from one wall and 9 inches from the other. What is the radius of the table?

51. Up the Garden Path

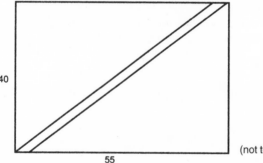

(not to scale)

The figure above shows a diagonal path across a rectangular garden measuring 55 yards by 40 yards. If the path is exactly 1 yard wide, what is its area?

52. Four Towns

There are four main towns in Puzzleland: A, B, C and D, lying at the corners of an imaginary 10-mile square. Their mayors are considering a joint project to link all four towns by road. The road must be as short as possible and still allow access between any one town and any other. The three designs below have been submitted:

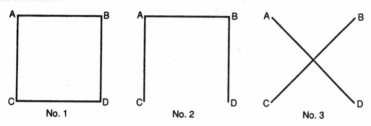

Design No. 1 requires 40 miles of roadway; No. 2 requires 30 miles; and No. 3 requires 28.3 miles. Is No. 3 the most economical design possible, or is there an even better solution?

53. Divide the Circles

You have three circles, A, B and C, with diameters of 5 centimeters, 4 centimeters and 3 centimeters respectively (see below).

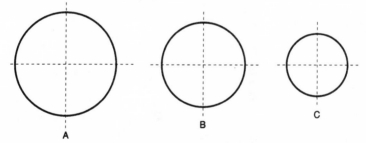

Divide the circles into 4 equal areas, using only a compass and a straight-edge.

54. The Spiral

How would you draw a spiral as shown above, using only a compass and a straight-edge?

55. The Banner of St. George

Looking at St. George's banner (below), consisting of a red cross on a white background, the following puzzle suggests itself.

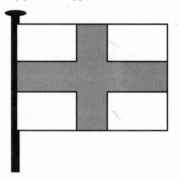

Suppose the banner measures 4 feet by 3 feet, what is the width of the cross if the white and red areas are to be equal?

2. Puzzles in Physics

56. Inner Space

Down in a deep coal mine, a mile or so below the surface of the Earth, is there more gravity or less gravity than there is at the entrance to the mine up at ground level? (Assume the Earth has a uniform density, and ignore any centrifugal and centripetal forces.)

57. Weigh-In

Even if you stand perfectly still on an accurate scale, the reading keeps oscillating around your average weight. Why?

58. The Rifle

A man holds a rifle horizontally 6 feet above the ground. At the moment he fires it, another bullet is dropped from the same height, 6 feet. Ignoring frictional effects and the curvature of the Earth, which bullet hits the ground first?

59. Cracked Up

Pour hot water into a thick drinking glass and into a thin wine glass. Which glass is more likely to crack?

60. Roll off

A ball, a disc and a ring, each 5 inches in diameter, sit at the top of an inclined plane. If all three objects start rolling down the incline at the same instant, which one will reach the bottom first? (Assume that they all roll efficiently—

that is, they are perfectly formed and don't wobble—and ignore any effect of air resistance and friction.)

61. Mountain Time

If you take your mechanical watch to the mountains, will it run faster or more slowly than usual?

62. Launch Pads

Why are space centers like Cape Canaveral usually located in tropical climates?

63. Poles Apart

Antarctica has 8 times as much ice as the Arctic. Why is there so much more ice at the South Pole than there is at the North?

64. A Hard Skate

Is it easier to ice skate when the air temperature is at 0°F or at 30°F?

65. The Suspended Egg

How would you make a raw egg float halfway between the surface and the bottom of a glass of water?

66. Two Bars of Iron

Two bars of iron lie on a table. They look identical, but one of them is magnetized (with a pole at each end), the other is not.

How can you discover which bar is magnetized if you are only allowed to shift them on the table, without raising them and without the help of any other object or instrument?

67. The Bird Cage

A cage with a bird in it, perched on a swing, weighs 4 lb. Is the weight of the cage less if the bird is flying about the cage instead of sitting on the swing? Ignoring the fact that if left in an airtight box for long the bird would die, would the answer be different if an airtight box were substituted for the cage?

68. Exception to the Rule

It is generally accepted that matter expands with increasing temperature, and contracts with decreasing temperature. There is one notable exception. Which is it, and why has nature provided for it?

69. Water Level Problem

An ice cube is floating in a beaker of water, with the entire system at 0° Centigrade (32°F). Just enough heat is applied to melt the ice cube without raising the temperature of the system. What happens to the water level in the beaker? Does it rise, fall, or stay the same?

70. Boat In the Bath

Rupert is sailing a plastic boat in his bathtub. The boat is loaded with nuts, bolts and washers. If Rupert dumps all these items into the water, allowing his boat to float empty, will the water level in the bath rise or fall?

71. Space Station

People have speculated that, one day in the far future, it may be possible to hollow out the interior of a large asteroid or moon and use it as a permanent space station. Assuming that such a hollowed-out asteroid is a perfect, non-rotating sphere with an outside shell of constant thickness, would an object inside, near the shell, be pulled by the shell's gravity field towards the shell or towards the center of the asteroid, or would it float permanently at the same location?

72. Bird on the Moon

Imagine a bird with a small, lightweight oxygen tank attached to its back so that it can breathe on the moon. Given that the pull of gravity is less than on the Earth, will the bird's flying speed on the moon be faster, slower, or the same as its speed on the Earth? Assume that, for the purposes of making the comparison, the bird has to carry the same equipment on Earth.

73. The Goldfish

A goldfish bowl, three-quarters full of water, is placed on a weighing scale. If a live goldfish is dropped into the water, the scale will show an increase in weight equal to the weight of the fish. However, assume that the goldfish is held by its tail so that all but the extreme tip of its tail is under water. Will this operation cause the scale to register a greater weight than it did before the fish was suspended?

74. Speed of Sound

The speed of sound in air is about 740 miles per hour. Suppose that a police car is sounding its siren and is driving towards you at 60 miles per hour. At what speed is the sound of the siren approaching you?

75. Two Bridges

Imagine two bridges that are exactly alike except that every dimension in one is twice as large as in the other. For example, the large bridge is two times longer, its structural members are two times thicker, and so forth. Which bridge is stronger, or is their strength the same?

76. Two Sailboats

Imagine two sailboats built to exactly the same proportions except that one is twice as large as the other: its masts are twice as thick; its sails are twice as long and twice as wide. Even though the sails are made out of the same kind of canvas, if the weight of the sail itself can be ignored, which sailboat will be more likely to have its sail torn by the force of the wind?

77. Aircraft Temperature

When an airliner is flying at an altitude of 30,000 feet, the temperature of the air outside may be as low as –30°F. One might think that this would require the use of heaters inside the cabin, but in fact an aircraft flying this high must use air-conditioners. Why?

78. Fan Power

You wanted to sail your model sailboat today, but there is no wind at all and your boat is in the doldrums. Would it be possible to propel your boat by mounting a battery-operated fan on the rear of the boat and directing the fan to blow wind into the sails?

79. Flags

Below are the flags of Pakistan (left) and Algeria (right).

The flags of the Comoro Islands, Mauritania and Tunisia have similar designs. Astronomically, is there anything "wrong" with them?

80. Where Are You?

Suppose you are a passenger in a doughnut-shaped space station. It is spinning around its hub to produce a simulated gravity of 1 g, exactly mimicking the gravity of Earth. You are in a small, windowless room, so you cannot see the rest of the space station. Inside your room everything seems "normal"— gravity seems to be operating on you exactly as it would on Earth. In fact, as far as your senses can tell you, you are on Earth.

You have a coin in your pocket. Is there a simple test you could do in your room that would confirm you are on a spinning space station and not on Earth?

81. Bridge Towers

The longest suspension bridge in the world is the Humber Estuary bridge in England, just short of 1 mile long between the supports. The two towers are not quite parallel. They are 1.375 inches farther apart at the tops than at the bottoms. Why?

82. Drops and Bubbles

If all space were empty except for two drops of water, the drops would be attracted to each other, according to Newton's Law of Gravity.

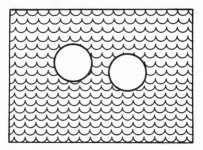

Now suppose all space were full of water except for two bubbles (see above). Would the bubbles move apart, towards each other, or not at all?

83. Special Sphere

If a sphere 20 feet in diameter enclosed a vacuum and weighed 300 lb., what amazing property would it possess?

84. Saving Energy

The boiling point of water is lower when atmospheric pressure decreases. Theoretically, then, you could attach a suction pump to a pot and suck out the air above the water level, and the water would come to a boil faster, thereby saving energy. Do you see anything wrong with this argument?

85. Shut That Door

You are in a room filled with 100 percent methane gas. What would happen if you struck a match?

3. Family Matters

86. The Telephone Call

A telephone conversation:

"Hello. Is this XYZ 8765?"

"Yes. Who's that?"

"What? You don't recognize my voice? Why, my mother is your mother's mother-in-law."

What is the relationship of the speakers?

87. Relations

"Jean is my niece," said Jack to his sister Jill.

"She is not my niece," said Jill.

Can you provide more than one possible explanation?

88. Sisters

Two look-alike girls sitting on a park bench are approached by a stranger. "You must be twins," he says.

The girls smile. "We have the same parents and were born on the same day in the same year, but, no, we're not twins."

How come?

89. How Close?

What is the closest relation that your mother's sister-in-law's brother-in-law could be to you?

90. The Painting

A man, looking at a painting, says to himself: "Brothers and sisters have I none, but that man's father is my father's son."

Who is the subject of the painting?

91. Sons' Ages

Two men, both the same age, sitting in a laundromat waiting for their washes to finish, struck up a conversation. One said to the other: "I have three sons. Let's see how many clues I have to give you before you're able to figure out their ages. To make it easy, we'll deal in whole years only. The question is: what are their ages next birthday?"

The other man agreed to give it a try.

The first man continued.

"One: the sum of my sons' ages is 13."

The second man shook his head.

"Two: the product of their ages is the same as your age."

"Carry on," said the second man.

"Three: my oldest son weighs 91 pounds. Four—?"

"Stop," said the second man. "I think I've got it."

What was his solution?

92. Four Families

Every Saturday, a gang of us play some kids from Brooklyn at baseball, but the other day, because of the flu, it looked like we'd have to miss a game. Then Chuck pointed out that, because some of us were better players than the rest, we'd still be able to pull together two evenly matched teams, though it would mean mixing up the two groups.

As I was getting ready to go to bat, who should walk up but the guy who's running for Mayor this year, saying we kids all looked so cute we could be brothers and sisters.

He patted my head for the benefit of the TV crew with him, so I told him we might look alike, but actually we belonged to four different families.

"There's my family," I explained. "Ours is the largest. Then there's the Browns, they're a smaller family. The Greens are smaller still, and the Black family is the smallest of all."

He wanted to know how many kids there were in each family, but I was in the mood to be difficult, so I told him, as it happens, you could multiply together the numbers of kids in each family and come out with the same number as was the number of fools who were going to vote for him this coming November. Then I told him how many votes I figured he'd get!

That should have been enough to persuade him to take a hike, but instead he said: "Well, let me see if I can work out the number of children in each family." After a moment, he said: "Tell me one thing, sonny: do the Blacks have more than one child?"

I didn't see how knowing that would help him much, so I answered. But I figured I'd outfox him by just saying Yes or No. Then the wiseguy looked straight at the TV cameras and announced how many kids each family had!

How did I answer his question, and what did the wiseguy say?

(For the baseball-ignorant: there are 9 players on a team.)

93. Father and Grandfather
Is it possible for one's grandfather to be younger than one's father?

94. Husbands and Fathers
Mary and Joan run into each other quite regularly at the local supermarket. One day, Mary mentioned: "You know, the other night I was suddenly struck by the odd way in which we're related. You realize, don't you, that our fathers are our husbands, as well as the fathers of our children?"

Explain.

95. How Many Children?
Each son in the Hubbard family has just as many brothers as sisters, but each daughter has twice as many brothers as sisters. How many boys and girls are there in the family?

96. Boys and Girls
A sultan wanted to increase the proportion of women to men in the population of his country so that the men could have larger harems. He proposed to ac-

complish this by passing a law which forbade a woman to have any more children as soon as she had given birth to a male child. Thus, he reasoned, some families would have several girls and only one boy, but no family could have more than one boy. After a time, females would outnumber males. Or would they?

(In the real world, slightly more than 50 percent of births are girls, but assume that, in the sultan's country, the natural birth ratio is exactly 50–50.)

97. Two Children

I have two children. At least one of them is a boy. What is the probability that both my children are boys?

My sister also has two children. The older child is a girl. What is the probability that both her children are girls?

98. Marital Problem

Jason and Dean were brothers. Jason married Jackie, and Dean married Denise. However, Jason and Denise have the same wedding anniversary. Dean's wedding anniversary was one month before this date, and Jackie's was one month after it. There have been no divorces or remarriages. How do you explain this?

4. Strange Situations

The problems that follow are rather different from usual puzzles in several respects.

First, although they can certainly be attempted solo, they are much more entertaining if tackled in a group: one person, who knows the solution, reads a puzzle aloud to the would-be solver, who asks questions to help him towards the answer. Effectively, this chapter can be used as the basis for a parlor game; the object of the game might be to solve the problems faster, or by asking fewer questions, or simply to solve more than the opposition.

While solving these puzzles is usually quite tough, creating new ones along similar lines need not be, once players have developed an understanding of what makes the puzzles tick. Devotees should therefore be able to keep the parlor game going after the examples I provide are exhausted.

Second, by their very nature, they provide the possibility of more than just one correct answer. Indeed, at first glance, they might seem to offer an almost infinite number of solutions, limited only by the breadth of the solver's imagination. Certainly, they do not have mathematically definitive answers in the way that conventional puzzles do but, in my experience on both sides of the fence, usually there is only one solution—ignoring minor variations of detail—that both participants find "right" or satisfactory. Substantially different alternatives tend to seem second-rate, even to the people who devised them, once the "official" answer is known.

Third, and most interestingly, their solution requires a radically different thought process from the standard puzzle. The issues in most conventional brainteasers are self-evident; they jump out at the puzzlist and trigger the thought processes, steering her or him in a direction (which may or may not turn out to be the right direction). For instance, if a problem requires calculation and paper and pen, the puzzlist is usually able to begin the figurework and put pen to paper as soon as he has finished reading it. That is not the case with the problems in this chapter. They are conundra in which half the battle is figuring out where to begin; which piece of given information is key to unravelling the mystery. They test a cross between lateral and intuitive thinking rather than mechanical intelligence, numeracy or acquired knowledge.

The chapter is called Strange Situations. Each problem is in the form of a narrative telling a story that seems out of kilter in some way. It may seem wildly improbable or contradictory; it may describe bizarre or otherwise inexplicable conduct; it may even seem to defy the laws of physics! But there is a rational explanation for the story, and the object of the puzzle is to find that explanation.

In this chapter, I admit to deviating somewhat from my general principle of avoiding unnecessary embellishment. In most of the narratives below, all the information provided is of some relevance, even if only tangential. But the occasional red herring has been allowed in to stimulate the reader's sense of bafflement (and, I hope, curiosity).

The ground rules are as follows: the solution obviously must fit all the given facts. It must conform to accepted norms of behavior and the physical world as we know it; in other words, you can take it for granted that none of the solutions involve supernatural powers.

Having said that, perhaps I should add one thing. In stating that there is a "rational explanation" for each of these strange situations, I do not mean to suggest that, once explained, they will all seem realistic in everyday-world terms; merely that they will suddenly make sense, according to the puzzlist's concept of acceptable logic. After all, the events described below did occur in Puzzleland!

99. The Severed Arm

A well-dressed man, let's call him John, enters a bar in the Bowery in Manhattan in autumn 1945. After looking around, he sits down next to a down-and-out-looking man, obviously the worse for drink. John strikes up a conversation and orders another round of drinks.

After some small talk, John makes the following proposition to the stranger: "I am willing to give you $20,000 in cash, and pay to have you fitted with an artificial limb, if you agree to have your left forearm amputated."

After some hesitation, the stranger agrees and they proceed to a dingy office in the Bowery, where John introduces the stranger to a third man, who performs the surgery.

Following the surgery, John packs the severed limb in dry ice and sends the parcel to an address in Los Angeles. At the same time, he sends cables to a number of addresses on the West Coast. Several days later eight men meet in LA. The parcel is opened, its contents inspected. The men express satisfaction and disperse.

Find an explanation to fit these facts.

100. The Elevator Stopped

A woman leaves her apartment, situated on the tenth floor of a high-rise building. She calls the elevator and begins to descend. The elevator comes to an abrupt stop between the fourth and third floors, and the light goes out. At that moment the woman's face turns ashen and she exclaims: "Oh God, my husband's dying."

Explain.

101. The Barber Shop

After leaving work one evening, David looks in on a barber shop on his route home. The barber shop is a one-man business, and, on that evening, the proprietor is shaving a customer and has a long line of other customers waiting their turn.

"How long will you be?" David asks. The barber, after a little reflection, replies, "Hour. Hour and a half."

David thanks him and leaves.

A few days later, David checks out the barber shop again. This time the barber estimates he will be able to get to David in about 40 minutes. Once again, David thanks him and leaves.

The following day, the barber is just finishing up with the last customer. "Give me 30 seconds," he tells David. David thanks the barber, but, instead of waiting, leaves the shop.

Explain.

102. The Car Crash

Harry Jones was driving home in his car, with his son Robert in the passenger seat. The car was involved in a head-on collision with a truck, killing Harry outright. Robert was seriously injured and taken to hospital by ambulance. In the hospital operating theater, his would-be surgeon took one look at Robert and said, "I'm sorry, but I can't operate on this patient—he's my son."

What is the explanation?

103. A Glass of Water

A man enters a bar and asks for a glass of water.

The bartender draws a gun and shoots into the ceiling.

The man thanks him and walks out.

Is the bartender crazy? Is the man? Or is this just another day in Dodge City?

104. The Unfaithful Wife

Chuck, a bestselling author of romantic fiction, had suspected for some time that his wife Eva was unfaithful, though he had no proof.

One afternoon, while Chuck was working on his latest bodice-ripper, Eva mentioned that she intended to go to the movies and would be out for a few hours. As Eva went to the door, Chuck looked at her pensively, then resumed his work.

Three hours later, Eva returned, took her coat off and asked Chuck whether he wanted some coffee. When she returned from the kitchen, Chuck asked her to sit down as he wanted to talk to her.

"Eva," he said, "I want a divorce."

Why?

105. The Suicide

The heiress to the Stanhope toothpaste fortune was found dead by her husband one morning, hanging from a chandelier in the master bathroom of their opulent townhouse. Her death had the police department mystified. Suicide, the initial theory, was ruled out because there seemed no way she could have hanged herself. There was no furniture directly beneath or close to the body, nothing that looked like it had been kicked away; the toilet was fifteen feet away in a corner; the bathtub and the Jacuzzi were both sunken into the floor. In short, the cops found nothing that the deceased might have used to stand on.

Murder was a possibility—she had many enemies but there were no signs of forced entry, and she was a recluse who never had visitors. Besides, her husband—the only other resident—claimed that when he returned home after spending the night at his parents, he'd found the bathroom door locked from the inside and had had to break it down.

Detective-Sergeant Plod had his suspicions about the husband, but nothing could be proved. The inquest handed down an open verdict, which enabled the husband to collect millions on his wife's life insurance.

Plod became obsessed with the case until his dying day, convinced that the husband must either have been involved, in some way, in the heiress's death, or at least have concealed evidence of her suicide before calling the police, thereby defrauding the insurance company. Was Plod right?

106. The Deadly Scotch

Mr. El and Mr. Lay went into a bar and each ordered a scotch on the rocks. Unknown to them, they both got drinks laced with poison. Mr. El downed his in one gulp and proceeded to chat for an hour while Mr. Lay drank his slowly. Later, Mr. Lay died, but Mr. El didn't. Why?

107. The Heir

The king dies and two men, the true heir and an impostor, both claim to be his long-lost son. Both fit the description of the rightful heir: about the right age, height, coloring and general appearance. Finally, one of the elders proposes a test to identify the true heir. One man agrees to the test while the other flatly refuses. The one who agreed is immediately sent on his way, and the one who refused is correctly identified as the rightful heir. Why?

108. Death in the Car

A man was shot to death while in his car. There were no powder marks on his clothing, which indicated that the gunman was outside the car. However, all the windows were up and the doors locked. After a close inspection was made, the only bullet-holes discovered were on the man's body. How was he murdered?

109. A Soldier's Dream

A soldier dreamed that his king would be assassinated on his first visit to a foreign city. He pleaded with the king to cancel the forthcoming trip, believing the dream might be a terrible omen. Pondering a moment, the king thanked the soldier for his advice and confirmed that he intended to take it, then ordered the soldier to be taken out and shot.

Why?

110. The Antique Candelabrum

The scene is a famous antique dealer's in London. A Rolls-Royce pulls up and a liveried chauffeur opens the door to a distinguished-looking elderly man, who enters the shop.

He points at a seventeenth-century candelabrum in the window. He examines it closely and then engages in an animated dialogue with the dealer. Eventually, he writes a check for $5,000 and departs with the candelabrum.

Shortly thereafter the dealer makes a number of telephone calls before closing his store. Two days later, he receives a call which clearly pleases him.

In the meantime, the distinguished-looking man has carefully wrapped the candelabrum he bought. A younger man, Robert, arrives at his suite at the Ritz, takes the wrapped candelabrum by taxi to the same antique dealer from whom the older man bought it. The dealer pays him $9,000 in cash for the candelabrum.

What happened?

111. The Two Accountants

Smith and Jones are partners in a small firm of accountants managing the investments of rock music stars. Jones is eight years Smith's senior, but is actually the junior partner, having joined the firm after Smith.

One morning, just before Christmas, Smith's secretary bursts into the conference room, where Smith and Jones are in a meeting, in search of her boss. "Your son just called," the secretary says. "Apparently, there's been a change of plan and you're going to your in-laws for the holiday."

On hearing this, Jones paled, got hold of a paperweight and threw it at his partner. Why?

112. The Judgment

A man is found guilty of first-degree murder. The judge says, "This is one of the most vicious criminal acts that has ever come before me and I am satisfied beyond any doubt that there are no mitigating factors. I wish I could impose the stiffest sentence at my disposal, but I have no choice but to let you go."

What is the reason for the judge's decision?

113. The Jilted Bride

In a mountain village in Switzerland, during the winter of 1693, a couple were being married. In the midst of the ceremony, a girl, jilted by the bridegroom, appeared and made a scene.

"The wedding bell will not ring," she said, before taking poison in front of the gathering.

Sure enough, when the bellringer tried, there was no sound from the bell. Most of the celebrants assumed she had bound the clapper in some way, though a few suggested that the girl was known to be a witch and had probably cast a spell.

After the ceremony, the bellringer climbed up to the belfry to investigate, but found everything in working order.

Was the girl a witch, after all?

114. The Elevator Rider

After reading about the death of jogging guru Jim Fixx, Bill eschewed all forms of exercise, considering it harmful to his health. He set up the International League of Couch-Potatoes and campaigned tirelessly on behalf of a sedentary lifestyle. His message soon struck a chord with a public tired of being harangued into doing exercise, to the point where business within the fitness industry began to suffer. Several of the country's largest health club chains and sports equipment manufacturers joined forces to combat the Couch-Potato movement; they spent lavishly on advertising, and put a private detective on Bill's tail in the hope of unearthing something to discredit him.

After six weeks of round-the-clock surveillance, the detective triumphantly submitted his report. While Bill seemed to follow his credo to the letter in public, the detective had noticed that when Bill took the elevator up to his apartment on the twenty-third floor of his building *alone,* he had a habit of getting out on the fifteenth floor and walking—one could almost call it *jogging*—up the last eight floors. He never did this, the report noted, if there were other people in the elevator. The detective had even managed to sneak a few photographs!

The story and the photos were leaked to the press, and Bill's reputation within couch-potato circles was ruined. He sued for libel and won. How come?

115. Crafty Cabby

In his book *Aha! Insight,* master gamesman Martin Gardner tells the story of a talkative, highly-strung woman who hailed a taxicab in New York City. During the journey the lady talked so much that the taxi driver got quite annoyed. He said, "I'm sorry, lady, but I can't hear a word you're saying. I'm deaf as a post, and the battery in my hearing aid is dead." This shut the woman up, but after she left the cab she figured out he had been lying to her. How did she know?

116. The Sharpshooter

A sharpshooter hung up his hat and put on a blindfold. He then walked 100 yards, turned around, and shot a bullet through his hat. The blindfold was a perfectly good one, completely blocking the man's vision. How did he manage this feat?

117. Death in Squaw Valley

A New York banker and his wife took their annual skiing holiday in the Valley. Late one afternoon, in bad visibility, the wife skidded over a precipice and broke her neck. The coroner returned a verdict of accidental death and released the body for burial.

In New York an airline clerk read about the accident. He contacted the police and gave them some information which resulted in the husband's arrest and indictment for first-degree murder. The clerk did not know the banker or his wife, and had never been to Squaw Valley.

What information did he give the police?

118. Burglars

Arthur lives with his parents in Chicago. Last week, while his parents were out, Arthur's next-door neighbor Sophie came round to spend the evening. At one point, she popped out to buy some cigarettes. Just then, two men burst into the apartment and, ignoring Arthur, took the TV set, the stereo and a computer.

Arthur had never seen the men before, and they had no legal right to remove the equipment, yet he did nothing to stop them. In fact, he didn't even act surprised by their behavior. Why not?

119. Insomnia

IBM executives held a sales conference at a hotel in Miami. Pete and Dave occupied adjoining rooms. After a strenuous day of presentations and partying, they went to their rooms. Despite being exhausted, Pete just could not get off to sleep. Eventually, at about two in the morning, he called the switchboard and asked to be put through to Dave's room. As soon as Dave picked up the phone, Pete replaced his and fell asleep.

Explain.

120. The North Pole

Porto Allegre in Brazil is situated on meridian 50° west and latitude 30° south, at a distance of approximately 15,000 kilometers from the North Pole. Eucla in Australia is also 15,000 kilometers from the North Pole.

What are the odds that Eucla, or for that matter any other spot on Earth equally distant from the North Pole, is more than 15,000 kilometers from Porto Allegre?

121. The Long Division

Solve the following sequence of divisions.

What is one-half of two-thirds of three-quarters of four-fifths of five-sixths of six-sevenths of seven-eighths of eight-ninths of nine-tenths of one hundred?

To find the solution without getting tied in knots you need a flash of inspiration.

122. Pool Resources

Brothers Andrew and Jim had just received their weekly pocket money, and planned to go to the fairground for Sunday afternoon. Jim, suspecting a smaller allowance, suggested to Andrew that they should pool and share their cash equally. "OK," said Andrew, "I know how much you have. If you guess how much I have I will split with you. To help you out, I will give you the following clue: if you give me $1, I shall have twice as much as you; if instead I give you $1, we shall each have the same amount."

How much pocket money has each brother received?

123. The Parking Dodge No. 1

Clive Gordon is a paragon of virtue. He does not smoke, does not drink, pays his taxes punctually and never exceeds the prescribed speed limits. There is however one quirk in this virtuous landscape of Clive's character. He hates traffic wardens and what they stand for and he considers all parking restrictions to be an abuse of power by oppressive and faceless authorities.

A great deal of his time is devoted to devising schemes which will enable him to park wherever he wants to and yet escape the scourge of parking tickets, clamping and being towed away.

One of his latest exploits is worth recording: a few months ago he took his Jaguar XJ6 to the continent and on his way to Vienna he stopped over in Zurich for a meeting with Dr. Hans Gruber, his Swiss lawyer. He parked his car in the Talstrasse and after a brisk 5 minutes walk arrived at Gruber's office. One hour later they had finished their business and when they shook hands Clive mentioned, by the way, where his car was parked.

Gruber (with a worried expression):

"You will be in serious trouble, parking in the Talstrasse is strictly limited to 20 minutes and penalties for exceeding it are very severe."

Gordon: "Don't worry, I will not be affected. However, in the unlikely event that I shall need your help, I will let you know."

What devilish scheme did he have in mind?

124. The Parking Dodge No. 2

After trying for weeks Clive Gordon managed to get two tickets for *Sunset Boulevard* at the Adelphi Theater on the Strand. Traffic was heavy and he arrived at the theater with only minutes to spare.

He had no time nor the inclination to look for a garage. He parked the car at a meter, which he did not even bother to feed, although it was already on excess.

He had no disabled badge and yet he knew he would not get a ticket. How did he manage it?

125. The Blip

After the invasion of Kuwait by Iraqi forces followed by Desert Storm, the coalition armies established their headquarters in Riyadh.

The basement of the Saudi airforce building served as intelligence gathering center monitored by American and British personnel recruited from the CIA and MI5.

Every morning between 5:30 A.M. and 6 A.M. two of them went to a small windowless room in the basement and waited for a transmission from one of their agents which the CIA had infiltrated into Baghdad.

At precisely 5:40 A.M. they heard a single blip, lasting less than half a second. Nevertheless the blip contained an important message causing great excitement.

Furthermore the recipients were satisfied that the information was genuine, not sent by an imposter and that their agent had not sent it under duress.

Can you explain this extraordinary phenomenon?

126. The Crossroad

George Genti was very proud of his new turbo coupe Rover as he drove north from Tunbridge towards Orpington. He was traveling at 50 miles per hour, well within the speed limit, and as he had just landed a profitable contract, he hadn't a care in the world. As he was just about to pass the Hadlow crossing, a black Nissan, traveling at high speed, collided with his Rover, severely damaging his back wing.

George, knowing he had right of way, was furious. He was on the point of blasting the driver of the Nissan with an effusion of four-letter words, when a gorgeous creature stepped out of the Nissan and, with the sweetest of smiles, accepted full responsibility and apologized profusely.

They moved their cars off the road, so as not to impede traffic and continued to chat. George was utterly captivated by the charm of this beautiful woman.

After exchanging first names she suggested that they celebrate their newly found friendship with a drop of the best. She produced a bottle of Grand Marnier and two glasses which she filled to the brim and toasted: "To us—bottoms up." George was happy to oblige, but was perplexed to see that she poured her own drink onto the ground.

Why did she behave so offensively?

127. A Wartime Mystery

The year was 1941 and the place the headquarters of the Red Army high command, facing the Kremlin.

It was one of the hottest June days on record in Moscow and, air-conditioning being out of action, all windows in the conference room on the tenth floor were wide open.

Gregory Topolev, Chief of Staff, was reporting to the assembled top brass on routine organizational matters, when suddenly the door opened to admit the head of the KGB, accompanied by a middle-aged civilian.

Topolev took one look at them, paled, and jumped out of the window. Considering that he knew both men very well and in fact considered them his friends, how do you explain his suicide?

128. The Black Forest

Susan and Tom were tracking the Black Forest, the highland in southwest Germany, when they discovered a young man hanging from a tree. Their first impulse was to cut him down, but as he was obviously dead they decided to report the find to the police.

On examination it was found that the man had died about 3 days before he was discovered, and that it was neither suicide nor murder.

What do you think had happened?

129. The Cabin in the Woods

There is an old German fairy-tale of two children, Hansel and Gretel, taking a stroll into the nearby woods and suddenly discovering a log cabin covered with mouth-watering chocolate cookies, the home of a wicked witch.

Our story is also about Hansel and Gretel, members of the village constabulary, who were ordered to search the forest which surrounded the neighborhood. They too came upon a cabin, but to their horror found it full of dead men, women and children. They did not die of natural causes and no crime was involved.

What do you think had happened?

130. The Sixpack

It was a hot, humid summer afternoon in East Finchley. George was carrying a sixpack of Coca-Cola, bought at the local supermarket, to stock up the fridge in his home more than a mile away.

The sixpack felt like a ton and although he changed the load from left to right hand in short intervals the burden became heavier by the minute.

Suddenly an idea struck him. If he were to drink three of the cans it would not only quench his thirst, but he would have much less to carry.

Do you agree with his reasoning?

131. The Gamblers

They knew it had to stop. Since his retirement Ian and Emma, his wife, had started to gamble. Initially it was out of boredom but before long they became addicted. Ian's poison was "Chemin de fer" and Emma loved "blackjack." Ian enjoyed a good pension and had substantial savings which enabled them to indulge in their favorite pastime not only in Las Vegas and Atlantic City but also in the fashionable European watering holes of Monte Carlo and Deauville.

One morning, having gambled all night, they made accounts and discovered that in a few months they had lost a fortune. They panicked and decided to join Gamblers Anonymous.

As a last fling they booked a suite near a casino for a seven-day stay. As fate would have it they won a five figure amount during the first three days. As it invariably happens, they made friends with gamblers at the same tables and in the intervals, over drinks, exchanged pleasantries and gambling anecdotes.

On the fourth day something happened, as a result of which Ian and Emma, and some of their new found friends, though not all, died a slow death. No crime or disease was involved.

What do you think had happened.

132. Spirit of St. Louis

Charles Lindbergh's dramatic solo flight from New York to Paris in May 1927 has become part of aviation history. The Spirit of St. Louis was a single-engined plane. The flight took $33\frac{1}{2}$ hours and the engine performed perfectly although it had not been tested nonstop for such a period of time. There was the risk of engine failure and the question you are asked to consider is this:

If the plane had been powered by two identical engines made by the same manufacturer and assuming that engineering technology was not sufficiently advanced to enable the plane to maintain flight on a single engine if one had failed, would Lindbergh have been safer or less safe with a twin-engine plane, or would it have made no difference?

133. The Appointment

Gerry had at last been offered a job as a barman in one of the most exclusive establishments in San Diego. He was to attend an interview with the manager at noon.

He set off from Los Angeles at 9 A.M., in ample time for the 124-mile trip, as he had decided to travel by car, which should not take more than two hours on Highway 101.

Although there was no breakdown, no accident, and traffic was light, he arrived three hours late for his appointment and lost the job.

Why did it take him five hours to travel from L.A. to San Diego?

5. General Puzzles

134. A Pair of Socks

Inside the drawer of a dressing-table in a dark room, there are 28 black socks and 28 brown socks. What is the minimum number of socks that I must take out of the drawer to guarantee that I have a matching pair?

135. Dubliners

Given that there are more people living in Dublin than there are hairs on the head of any Dubliner, and that no Dubliner is totally bald, does it necessarily follow that there must be at least two Dubliners with exactly the same number of hairs?

Here is a variant of the same problem: on the island of Alopecia, the following facts are true:

1. No two islanders have exactly the same number of hairs.
2. No islander has precisely 450 hairs.
3. There are more islanders than there are hairs on the head of any one islander.

What is the largest possible number of islanders on Alopecia?

136. The Clock-watcher

George did not have a wristwatch, but he did have an accurate clock. However, he sometimes forgot to wind his clock. Once when this happened, George went to his sister's house, spent the evening with her, went back home and set his clock. How could he do this without knowing beforehand the length of the trip?

137. The Prisoners' Test

A wicked king amuses himself by putting 3 prisoners to a test. He takes 3 hats from a box containing 5 hats—3 red hats and 2 white hats. He puts one hat on each prisoner, leaving the remaining hats in the box. He informs the men of the total number of hats of each color, then says, "I want you men to try to determine the color of the hat on your own head. The first man who does so correctly, and can explain his reasoning, will immediately be set free. But if any of you answers incorrectly, you will be executed on the spot."

The first man looks at the other two, and says, "I don't know."

The second man looks at the hats on the first and third man, and finally says, "I don't know the color of my hat, either."

The third man is at something of a disadvantage. He is blind. But he is also clever. He thinks for a few seconds and then announces, correctly, the color of his hat.

What color hat is the blind man wearing? How did he know?

138. Above or Below

The first 25 letters of the alphabet are written out as shown below—with some letters above the line and some below. Where should Z go: above the line or below, and why?

A	EF	HI	KLMN		T	VWXY
BCD	G	J		OPQRS	U	

139. Strange Symbols

When Alice first stepped through the looking glass, she saw this strange set of symbols on a sign:

"What does it mean?" she wondered. "It looks like a secret code or the alphabet of some foreign language." Alice studied the sign a little longer and then had a thought. "It's a sequence with a definite pattern." When you recognize the pattern you should have no difficulty in drawing the next figure in the sequence. What is the next symbol?

140. Product

What is the product of the following series?

$$(x - a)(x - b)(x - c) \ldots (x - z)$$

141. What Are They?

"How much does one cost?" asked the customer in a hardware store.

"Twenty cents," replied the clerk.

"And how much will twelve cost?"

"Forty cents."

"OK. I'll take nine hundred and twelve."

"Fine. That will be sixty cents."

What was the customer buying?

142. Mending the Chain

Alice has 4 pieces of gold chain, each consisting of 3 links. She wants to have the pieces joined together to make a necklace, but she is afraid she can't afford it.

The jeweler eyes the 4 pieces of chain on his workbench. "I charge $1 to break a link and $1 to melt it together again. To fit the pieces together, I'll have to break and rejoin, four links. That will be $8."

Alice knew she had less than $7. "I don't have enough money," she said sadly. "I was hoping to wear the necklace tonight, but I guess that's out of the question." Alice gathered the pieces of chain and prepared to leave the store.

Just then the jeweler said, "Wait, I've thought of another way." Sure enough, he had. How did he do it, and how much did he charge?

143. Fast Fly

A Ferrari is traveling at 30 miles per hour on a head-on collision course with a Maserati, which is being driven at a leisurely 20 miles per hour. When the two cars are exactly 50 miles apart, a very fast fly leaves the front fender of the Ferrari and travels towards the Maserati at 100 miles per hour. When it reaches the Maserati, it instantly reverses direction and flies back to the Ferrari and continues winging back and forth between the rapidly approaching cars. At the moment the two cars collide, what is the total distance the fly has covered?

144. How Fast?

The Baja Road Race is 1,000 miles long. At the half-way point, Speedy Gonzales calculates that he has been driving at an average speed of 50 miles per hour. How fast should he drive the second half of the race if he wants to attain an overall average of 100 miles per hour?

145. Which Coffeepot?

A camper is trying to decide which of two coffeepots (pictured below) to buy. He wants the one that will hold the most coffee.

Which one do you recommend? (Assume that the cross sections of the two coffeepots are exactly the same.)

146. Cocktail

Four toothpicks and a penny are arranged as shown above, to represent a cherry in an old-fashioned glass. Move 2 toothpicks and get the cherry outside of the glass. The glass may have any position at the end, but the cherry cannot be moved.

147. The Explorer and the Bear

You may have come across this well-known riddle.

An explorer walked 1 mile due south, turned and walked 1 mile due east, turned again and walked 1 mile due north. He found himself back where he started, and was then eaten by a bear. Question: What color was the bear?

The answer is "White," because he must have started at the North Pole, so the bear that ate him must have been a polar bear.

However, the North Pole is not the only starting point that satisfies the conditions of the riddle. From what other point on the globe can you walk 1 mile south 1 mile east, 1 mile north and find yourself at your original location?

148. Buttons and Boxes

Imagine you have 3 boxes, one containing 2 red buttons, one containing 2 green buttons, and the third, 1 red button and 1 green button. The boxes are labeled according to their contents—RR, GG and RG. However, the labels have been switched so that each box is now incorrectly labeled. Without looking inside, you may take 1 button at a time out of any box. Using this process of sampling, what is the smallest number of buttons needed to determine the contents of all 3 boxes?

149. Manhattan and Yonkers

Sharon lives in Riverdale, near the train station. She has two boyfriends, Wayne in Yonkers and Kevin in midtown Manhattan. To visit Wayne, she takes a train from the northbound platform; to visit Kevin she takes a train from the southbound platform. Since Sharon likes both boys equally well, she simply gets on the first train that comes along. In this way she lets chance determine whether she goes to Yonkers or Grand Central. Sharon reaches the station at a random moment every Saturday afternoon. The northbound and southbound trains arrive at the station equally often—every 10 minutes. Yet, for some reason, she finds herself spending most of her time with Wayne in Yonkers. In fact, on average, she goes there 9 times out of 10. Can you explain why the odds are so heavily in Wayne's favor?

150. Counterfeit Coins

In the figure below are 10 stacks of coins, each consisting of 10 silver dollars. One entire stack is counterfeit, but you do not know which one. You do know the weight of a genuine silver dollar, and you are also told that each counterfeit coin weighs 1 gram more or less than a genuine coin. You have a pointer scale which you use to weigh the coins.

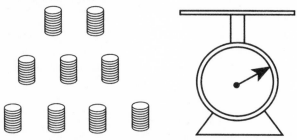

What is the smallest number of weighings needed to determine which of the stacks is counterfeit?

151. Fake!

Here is another counterfeit coin puzzle.

One of 12 identical-looking coins is counterfeit. The weight of the counterfeit coin is different from that of the genuine coins. Using only a simple balance, how can the counterfeit coin be identified in just 3 weighings?

152. How Long?

A bolt of cloth is colored as follows: one-third and one-quarter of it are black, the other 8 yards are gray. How long is the bolt?

153. Arithmetic Problem

In order to encourage his son to study arithmetic, a father agrees to pay his boy 8 cents for every problem correctly solved and to fine him 5 cents for each incorrect solution. At the end of 26 problems, neither owes anything to the other. How many problems did the boy solve correctly?

154. The Cloak

A butler is promised $100 and a cloak as his wages for a year. After 7 months he leaves this service, and receives the cloak and $20 as his due. How much is the cloak worth?

155. Water Lentils

The water lentil reproduces by dividing into two every day. Thus on the first day we have 1, on the second day we have 2, on the third 4, on the fourth 8, and so on. If, starting with one lentil, it takes 30 days to cover a certain area, how long will it take to cover the same area if we start with 2 lentils?

156. Two Steamers

Two steamers simultaneously leave New York for Lisbon, where they spend 5 days before returning to New York. The first makes 30 miles an hour going and 40 miles an hour returning. The second makes 35 miles an hour each way. Which steamer gets back first?

157. Unequal Scales

Discovering his scales are faulty, a grocer resolves to weigh all his customers' orders in two halves, putting the first half in the left-hand pan and weights in the right, then reversing the procedure. This, he believes, will be fair to both his customers and himself. Is he right?

158. The Long Division

(This type of puzzle is called a cryptogram.)

I was sitting before my chessboard pondering a combination of moves. At my side were my son, aged 8, and my daughter, 4 years old. The boy was busy with his homework, which consisted of some exercises in long division, but he was rather handicapped by his mischievous sister, who kept covering up his figures with chessmen. As I looked up, only two digits remained visible (see below).

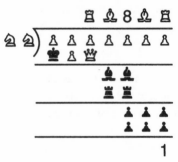

Can you calculate the missing digits without removing the chessmen?

159. Two Bolts

Below are two identical bolts held together with their threads in mesh. While holding bolt A stationary, you swing bolt B around it counterclockwise, as shown.

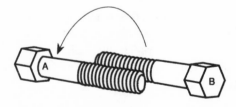

Will the bolt heads get nearer, move farther apart, or stay the same distance from each other? What will happen if you swing bolt B the opposite way—in a clockwise direction around A?

160. The Duel

Lord Montcrief, Sir Henry Darlington and the Baron of Rockall decided to resolve a quarrel by fighting an unusual sort of pistol duel. They drew lots to determine who fired first, who second and who third. They then took their places at the corners of an equilateral triangle. It was agreed that they would fire single shots in turn and continue in the same order until two of them were dead. The man whose turn it was to fire could aim at either of the other two. It was known that Lord Montcrief always hit his target, that Sir Henry was 80 percent accurate and that the Baron was only 50 percent accurate.

Assuming that all three adopted the best strategy, and that no one was killed by a wild shot not intended for him, which one had the best chance of surviving?

161. Fair Shares

"Der Eine dielet, der Andere kieset."

This is an old German saying which prescribes how to divide a cake between two people and ensure that both of them are satisfied. The translation is: "One divides, the other chooses."

Can you devise a procedure so that n persons can divide a cake between them and each one is satisfied that he has obtained at least $\frac{1}{n}$ of the cake?

162. Racing Driver

A racing driver drove around a 6-mile track at 140 miles per hour for 3 miles, 168 miles per hour for 1.5 miles, and 210 miles per hour for 1.5 miles. What was his average speed for the entire 6 miles?

163. The Side View

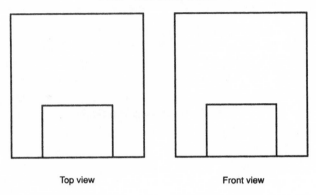

Top view   Front view

Look at the above two views of an object. What is the side view like? A perspective view?

164. The Striking Clock

It takes a clock two seconds to strike 2 o'clock. How long will it take to strike 3 o'clock?

165. Red or Green

You are given 4 pieces of cardboard and told that each one is either red or green on one side, and that each one has either a circle or a square on the other side. They appear on the table as follows:

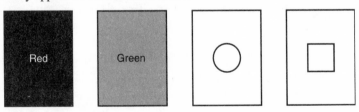

How many cards, and which ones, must you turn over in order to have sufficient information to answer the question: Does every red card have a square on its other side?

166. The River

A man takes his motor boat down a river to his pub. Going with the current he can cover the 2 km in 4 minutes. Returning against the current, which is steady, it takes him 8 minutes. How long does it take him at still water, when there is no current?

167. Changing the Odds

In a distant kingdom lived a king who had a beautiful daughter. She fell in love with a humble peasant boy, whom she wanted to marry. The king, who had no intention of consenting to the marriage, suggested that the decision be left to chance. The three were standing in the castle's forecourt, which was covered with innumerable white and black pebbles. The king claimed to have picked up one of each color, and put the two pebbles into his hat. The suitor had to pick one pebble from the hat. A white pebble meant that he could marry the king's daughter; a black, that he was never to see her again. The peasant boy was poor, but not stupid. He noticed that by sleight of hand, the king had put two black pebbles into the hat.

How did the suitor resolve his predicament without calling the king a cheat?

168. Zeno's Paradox

In the fifth century B.C., Zeno, using his knowledge of infinity, sequences and partial sums, developed this famous paradox. He proposed that, in a race with Achilles, the tortoise be given a head start of 1,000 meters. Assume that Achilles could run 10 times faster than the tortoise. When the race started and Achilles had gone 1,000 meters, the tortoise would still be 100 meters ahead. When Achilles had gone the next 100 meters the tortoise would be 10 meters ahead.

Zeno argued that Achilles would continually gain on the tortoise, but he would never reach him. Was his reasoning correct? If Achilles were to pass the tortoise, at what point of the race would it be?

169. The Boy and the Girl

A boy and a girl are talking.

"I'm a boy," says the one with black hair.

"I'm a girl," says the one with red hair.
If at least one of them is lying, which is which?

170. Infinity and Limits

The figure below illustrates circumscribing regular polygons. The number of sides of the polygon successively increases. Will the radii grow without limit? If not, can you estimate the limit, assuming the smallest circle to have a radius of 1 inch?

171. Hats in the Wind

Ten people, all wearing hats, were walking along a street when a sudden wind blew their hats off. A helpful boy retrieved them and, without asking which hat belonged to which person, handed each person a hat. What is the probability that exactly 9 of the people received their own hats?

172. Handshakes

Is the number of people in the world who have shaken hands with an odd number of people odd or even?

173. Move One

The following equation, in Roman numerals, suggests that 6 plus 2 equal 5. Can you correct it by moving only one line?

174. A Chiming Clock

A clock chimes every hour on the hour, and once each quarter hour in between. If you hear it chime once, what is the longest you may have to wait to be sure what time it is?

175. Ring Around the Circle

The three circles below are all the same size. How many circles will it take in all to make a complete ring around the shaded circle? Do this one without using coins or other circles, and prove your answer.

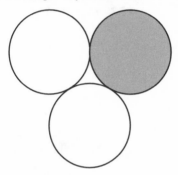

176. A Unique Number

The following number is the only one of its kind. Can you figure out what is so special about it?

8,549,176,320

177. The Boss and His Chauffeur

A chauffeur always arrives at the train station at exactly 5 o'clock to pick up his boss and drive him home. One day, his boss arrives an hour early, starts walking home, and is eventually picked up. He arrives home 20 minutes earlier than usual. How long did he walk before he met his chauffeur?

178. Weather Report

"Can you tell me what the temperature has been at noon for the past five days?" John asked the weatherman.

"I don't exactly recall," replied the weatherman, "but I do remember that the temperature was different each day, and that the product of the temperatures is 12."

Assuming that the temperatures are expressed to the nearest degree, what were the five temperatures?

179. Can You?

"We eat what we can, and we can what we can't."

Can you explain who could make this statement.

180. Presidents

"Here is an odd item, Professor Flugel," said Tom, looking up from his newspaper. "It says here that three of the first five presidents of the United States died on the Fourth of July. I wonder what the odds are against a coincidence like that."

"I'm not sure," replied the professor, "but I'm willing to give 10 to 1 odds I can name one of the three who died on that date."

Assuming that the professor had no prior knowledge of the dates on which any of the presidents died, was he justified in offering such odds?

181. Jim and George

How is it possible for Jim to stand behind George and George to stand behind Jim at the same time?

182. Two Sizes of Apples

A man had an apple stall and sold his larger apples at 3 for $1 and his smaller apples at 5 for $1. When he had just 30 apples of each size left to sell, he asked his son to look after the stall while he had lunch. When he came back from lunch the apples were all gone and the son gave his father $15. The father questioned his son. "You should have received $10 for the 30 large apples and

$6 for the 30 small apples, making $16 in all." The son looked surprised. "I am sure I gave you all the money I received and I counted the change most carefully. It was difficult to manage without you here, and, as there was an equal number of each sized apple left, I sold them all at the average price of 4 for $1. Four into 60 goes 15 times so I am sure $15 is correct.

Where did the $1 go?

183. The Sequence

A friend asks you to continue the following sequence.

OTTFFS . . .

When this suggests nothing to you he adds another term:

OTTFFSS . . .

The pairs of letters TT, FF, SS now suggest something, but you still cannot deduce the sequence. He adds another term:

OTTFFSSE . . .

You are still worried by the initial term O, but otherwise every other pair of terms seems to run in reverse sequence through the alphabet (TT, SS, RR, QQ . . .) and (FF, EE, DD, CC . . .), so you write down

OTTFFSSEERRDD . . .

What is the real solution?

184. The Half-full Barrel

Two farmers were staring into a large barrel partly filled with ale. One of them said: "It's over half full!" But the other declared: "It's more than half empty." How could they tell without using a ruler, string, bottles, or other measuring devices if it was more or less than exactly half full?

185. The Marksman

One marksman can fire 5 shots in 5 seconds while another can get off 10 shots in 10 seconds. (We will assume that timing starts when the first shot is fired and ends with the last shot, but the shots themselves will be assumed to take no time.)

Which man can fire 12 shots in a shorter time?

186. The Square Table

A square table is constructed with an obstruction in the middle of it, so that when 4 people are seated, one on each side, each can see his neighbors to right and left but not the person seated directly across. The 4 people seated at this table were told to raise their hands if, when looking to the right and left, they saw at least one woman. They were also told to announce the sex of the one person whom they could not see, if they could figure it out.

Since, as a matter of fact, all 4 people were women, each raised her hand, but then several minutes went by before one of them announced that she was certain that the person seated opposite her was a woman.

How could she logically have come to that conclusion?

187. The Marbles

Two bags each contain 3 red, 3 white, and 3 blue marbles. Without looking, someone removes from the first bag the largest number of marbles that it is possible to remove while still being sure that at least one of each color remains. These marbles are put into the second bag. Now he transfers back (without looking, of course) the smallest possible number of marbles that will assure there being at least 2 of each color in the first bag.

How many marbles remain in the second bag?

188. A Carbon Copy

For some reason, probably dishonest, someone wants to write a letter that will appear to be a carbon copy. He has only one sheet of paper and one piece of carbon paper, which he places under the letter paper with the carbon side facing the back of the sheet.

He then writes with a pen which has no ink so the writing will appear only on the underside of the paper. If he wrote normally on the top of the paper (with the inkless pen) the writing would appear inverted on the underside, so he decides he will have to write invertedly. For instance, instead of the usual F, he will write Ⅎ; instead of R, he will write Я.

Should he start the letter in the upper left-hand corner, upper right, lower right, or lower left?

189. Walking in Step

A man and a woman, walking together, both start out by taking a step with the left foot. In order to keep together, the man, whose stride is longer, takes two steps while the woman takes three.

How many steps will the woman have taken when they are both about to step out together with their right feet for the first time?

190. Razor Shortage

Due to rumors that there is a world shortage, people have started hoarding packs of razors. By the time I reached the supermarket, there were no packs left. However, the two people who checked out ahead of me had bought 3 packs and 5 packs respectively, and they offered to share them with me so that we each took home the same number of razors. I paid them $8.

How did they divide the $8?

191. Unusual Equations

$$5 + 5 + 5 = 550$$

a) Using a line equal to a hyphen, i.e. a - in any position, rectify the above equation.
b) Using four nines, make them equal 100.
c) Do the same with four sevens.
d) Arrange three nines to equal 20.

192. The d'Alembert Paradox

The French mathematician d'Alembert (1717–83) considered the probability of throwing heads at least once when tossing a coin twice. "There are only three possible cases," he argued, "(a) Tails appears on the first toss and again on the second toss. (b) Tails appears on the first toss and heads on the second toss. (c) Heads appears on the first toss (therefore in this case there is no longer need to carry out the second toss)."

"It is quite simple," he stated, "because there are only three possible cases; and as two of these are favorable, the probability is 2/3." Was his reasoning correct?

193. Jugs

There are many variations of puzzles involving decanting (pouring from one container to another). The one presented here is known to be at least 400 years old.

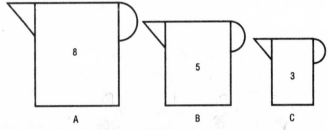

Three jugs (above) have capacity for 8, 5 and 3 pints respectively. The 8-pint jug is filled entirely with water and the other two jugs are empty. Your task is, by decanting, to divide the water into two equal parts, i.e. 4 pints in jug A and 4 pints in jug B, leaving the smallest jug empty. None of the jugs is calibrated, so the only way the task can be successfully performed is to pour water from one jug to another until the first jug is entirely empty or the second jug is entirely full. You must assume that the decanting is done with great care so that no water is spilled.

What is the least number of decantings in which the task can be achieved?

194. One-two-three

I first heard this puzzle from a Mensa member who had come across it years before at a mathematical conference in Holland. I gave the puzzle to a friend of mine who could not solve it and took it along to his club. The members there pondered over it for many hours without success. Finally, one member took it home to show his 12-year-old son, who solved it in 5 minutes. It is that type of puzzle.

What is the next row of digits?

```
1
1  1
2  1
1  2  1  1
1  1  1  2  2  1
3  1  2  2  1  1
1  3  1  1  2  2  2  1
1  1  1  3  2  1  3  2  1  1
?  ?  ?  ?  ?  ?  ?  ?  ?  ?
```

195. A Bottle of Wine

A bottle of wine costs $10. If the wine is worth $9 more than the bottle, what is the value of the bottle?

196. Roll-a-penny

We all remember the old fairground game which is still to be found, "Roll-a-Penny," in which the penny has to come to rest clear of any of the crossed lines.

How does one calculate the chances of winning?

Here is an example of a typical linoleum-topped table on which the game is played. How much better are the odds for the "house" than for the punter?

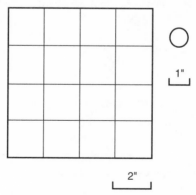

197. The Missing £

This is a very old English puzzle. Three men dined in a restaurant and the bill came to £25.

Each man gave the waiter £10 and told him to keep £2 out of the change as a tip.

The waiter returned with £3 and proceeded to give each man £1.

The meal had therefore cost £27 plus £2 for the waiter's tip. Where had the other £1 gone?

198. Gold Card

This puzzle was a favorite of Wild West card-sharps, who won thousands of dollars from the unsuspecting gold prospectors. In a saloon, the gambler would gather a crowd and place 3 cards into a hat. These cards were colored:

GOLD on one side, GOLD on reverse;
SILVER on one side, SILVER on reverse;
GOLD on one side, SILVER on reverse.

Reverse Gold   Reverse Silver   Reverse Silver

The gambler then asked an onlooker to draw a card from the hat and place it on the table, thus:

Then the gambler said: "The reverse side is either gold or silver as the card cannot be the silver/silver card. Therefore it is either the gold/silver card or the gold/gold card, an even chance! I will bet even money one dollar that the reverse side is gold."

Is this a fair bet?

199. The Airplane

An airplane flies in a straight line from airport A to airport B, then back in a straight line from B to A. It travels with a constant engine speed and there is no wind. Will its travel time for the same round trip be greater, less or the same if, throughout both flights, at the same engine speed, a constant wind blows from A to B?

200. Skiing the Atlantic

The members of a water-skiing club, based in Long Beach, want to try the first circumnavigation of the globe, via the Capes and the Timor Sea, by a water skier. Aside from all other difficulties associated with this feat, they have a problem in that the best speedboats can only carry enough fuel to get half-way around—though they are able to transfer fuel from one boat to another without stopping.

The task the club has set itself is to find a way to get one speedboat and a skier all the way around the Earth without stopping, while, of course, ensuring that all the support speedboats (carrying fuel supplies) return safely to Long Beach. Is this possible?

201. Four Bugs

Four bugs—A, B, C and D—occupy the corners of a 10-inch square (see below). A and C are male, B and D are female. Simultaneously, A crawls directly towards B, B towards C, C towards D and D towards A. If all four bugs crawl at the same constant rate, they will describe four congruent logarithmic spirals which meet at the center of the square.

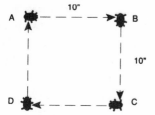

How far does each bug travel before they meet? (The problem can be solved without calculus.)

202. Birth Dates

You are one of 24 guests at a cocktail party and you bet a friend that there is at least one coincidence of birth dates among the people in the room. If you exclude leap years do you have a better chance of winning or losing the bet?

203. Changing Money

Here is that very puzzling story having to do with foreign exchange. The governments of two neighboring countries—let's call them Eastland and Westland—had an agreement whereby an Eastland dollar was worth a dollar in Westland, and vice versa. But, one day, the government of Eastland decreed that thereafter a Westland dollar was to be worth only 90 cents in Eastland. The next day the Westland government, not to be outdone, decreed that thereafter an Eastland dollar was to be worth only 90 cents in Westland.

A young entrepreneur named Malcolm lived in a town which straddled the border between the two countries. He went into a store on the Eastland side, bought a 10-cent razor, and paid for it with an Eastland dollar. He was given a Westland dollar, worth 90 cents there, in change. He then crossed the street, went into a Westland store, bought a 10-cent package of blades, and paid for it with the Westland dollar. There he was given an Eastland dollar in change. When Malcolm returned home, he had his original dollar and his purchases. And each of the tradesmen had 10 cents in his cash-drawer.

Who, then, paid for the razor and blades?

204. Rice Paper

Suppose you have a large sheet of very thin rice paper one-thousandth of an inch thick, or 1,000 sheets to the inch. You tear the paper in half and pile the two pieces on top of each other. You tear the two piled sheets in half and stack them again, then a third time so that you now have a stack of 8 sheets, and so on. If you keep doing that 50 times, how tall will the final stack be? The usual responses are amusing. Some people suggest a foot, others go as high as several feet, a few even suggest a mile. What do you think?

205. Two Discs

Consider the two equal circular discs, A and B, in the figure below.

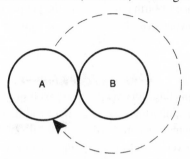

If B is kept fixed and A is rolled around B without slipping, how many revolutions will A have made about its own center when it is back in its original position?

206. The Slab and the Rollers

If the circumference of each roller in the figure below is 1 foot, how far forward will the slab have moved when the rollers have made one revolution?

207. The Broken Stick

If a stick is broken at random into 3 pieces, what is the probability that the pieces can be put together to make a triangle?

208. The Slow Horses

An aged and, it appears, somewhat eccentric king wants to pass his throne on to one of his 2 sons. He decrees that his sons will race their horses and that the son who owns the slower horse shall become king. The sons, each fearing that the other will cheat by having his horse go less fast than it is capable of, ask a wise man's advice. With just two words, the wise man ensures that the race will be fair.

What does he say?

209. Hourglass

You have two hourglasses—a 4-minute glass and a 7-minute glass. You want to measure 9 minutes.

How do you do it?

210. Sorting the Numbers

The numbers from 0 to 14 are divided into three groups as follows:

0 3 6	1 4 7	2 5 10
8 9	11 14	12 13

Group 1   *Group 2*   *Group 3*

Which groups do the next three numbers—15, 16 and 17—belong in?

211. Categories

The letters of the alphabet can be arranged in four distinct groups. The first 13 letters establish the categories:

    A M
    B C D E K
    F G J L
    H I

Place the remaining 13 letters in their proper categories.

212. What Weights?

A boy selling fruit has only three weights, but with them he can weigh any whole number of pounds from 1 lb. to 13 lbs. inclusive. What weights does he have?

213. Rope Trick

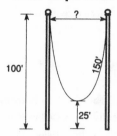

Two flagstaffs are each 100 feet high. A rope 150 feet long is strung between the tops of the flagstaffs. At its lowest point the rope sags to within 25 feet of the ground (see figure above). How far apart are the flagstaffs?

214. Catching the Bus

Juliette and her sister Lucille live together in the town of Montreux in the Swiss Alps. In the springtime, one of their favorite walks is to go up to the lovely fields of narcissi growing on the mountain slopes nearby.

On one occasion they came to a long straight stretch of road, and, at a certain point on it, they left the road and walked at right angles across a field to a large clump of narcissi. Juliette stopped to pick some of the flowers 40 meters away from the road, while Lucille also collected some flowers another meter farther away. Suddenly they looked up to see a bus going along the road to Montreux. When they had decided to ride home, the bus was 70 meters away from the point where they left the road to walk across the field.

They ran at half the speed the bus traveled to the point where they left the road and missed the bus! There is at least one point on that stretch of road where the bus could have been caught.

Can you calculate where they should have run and if both of the sisters could have caught the bus?

215. Bus Timetable

A man drives along a main road on which a regular service of buses is in operation. He is driving at a constant speed. He notices that every 3 minutes he meets a bus and that every 6 minutes a bus overtakes him. How often does a bus leave the terminal station at one end of the route?

216. How Many Hops?

You are standing at the center of a circle of radius 9 feet. You begin to hop in a straight line to the circumference. Your first hop is $4\frac{1}{2}$ feet, your second $2\frac{1}{4}$ feet, and you continue to hop each time half the length of your previous hop. How many hops will you make before you get out of the circle?

217. Decaffeiné

This is to be solved in the head, without paper and pencil.

If some coffee is "97 percent caffeine-free," how many cups of it would one have to drink to get the amount of caffeine in a cup of regular coffee?

218. Speed Test

Complete this equation in less than one minute:

$$\frac{1234567890}{1234567891^2 - (1234567890 \times 1234567892)} = ?$$

219. Walking Home

John was going home from Brighton. He went halfway by train, 15 times as fast as he goes on foot. The second half he went by ox team. He can walk twice as fast as that.

Would he have saved time if he had gone all the way on foot? If so, how much?

220. The Watchmaker

A watchmaker was telephoned to make an urgent house call to replace the broken hands of a clock. He was sick, so he sent his apprentice.

The apprentice was thorough. When he finished inspecting the clock it was dark. He hurriedly attached the new hands, but mixed up the hour and the minute hands. He then set the clock by his pocket watch. It was 6 o'clock, so he set the big hand at 6 and the little hand at 12.

The apprentice returned, but soon the telephone rang. He picked up the receiver only to hear the client's angry voice: "You didn't do the job right. The clock shows the wrong time."

Surprised, he hurried back to the client's house. He found the clock showing not much past 8. He handed his watch to the client, saying: "Check the time, please. Your clock is not off even by 1 second."

The client had to agree.

Early next morning the client telephoned to say that the clock hands, having apparently gone berserk, were moving around the clock at will. When the apprentice rushed over, the clock showed a little past 7. After checking with his watch, the apprentice got angry:

"You are making fun of me! Your clock shows the right time!"

Do you know what was going on?

221. The Lead Plate

The builders of an irrigation canal needed a lead plate of a certain size, but had no lead in stock. They decided to melt some lead shot. But how could they find its volume beforehand?

One suggestion was to measure a ball, apply the formula for the volume of a sphere, and multiply by the number of balls. But this would take too long, and anyway the shot wasn't all the same size.

Another was to weigh all the shot and divide by the specific gravity of lead. Unfortunately, no one could remember this ratio, and there was no manual on the site.

Another was to pour the shot into a gallon jug. But the volume of the jug is greater than the volume of the shot by an undetermined amount, since the shot cannot be packed solid and part of the jug contains air.

Do you have a suggestion?

222. The Caliper

A student had to measure a cylindrical machine part with indentations at its bases (see overleaf).

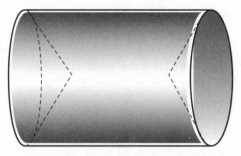

He had no depth gauge, only a caliper and a ruler. The problem was, he could find the distance between the indentations with the caliper, but he would have to remove the caliper to measure its spread on the ruler. But to remove the caliper he would have to open the legs, and then there would be nothing to measure.

What did he do?

223. Fuel Tanks

A small fuel station supplies the farms on the outskirts of town with fuel for their vehicles. It receives deliveries once a month, and stores the fuel in 6 tanks, of different sizes, reserved exclusively for the farms. Five tanks hold diesel fuel and the sixth a special blend of unleaded gasoline and alcohol. The tanks are labeled to show their capacities as follows:

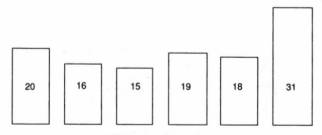

Some of the diesel is sold to Joe Smith, and the balance—exactly twice as much as Joe takes—is bought by Pete Brown's place. The unleaded/alcohol blend is for Dave Jones. Last month, all three farmers turned up together to collect their supplies, at a time when the only person working at the gas station was a new kid who hadn't yet been told which tank contained which type of fuel. Neither he nor the farmers had any means of measuring the fuel, the tanks in which the farmers carried off their supplies having larger capacities than the amount of fuel they bought each month, and not being calibrated in any way.

Nevertheless, when the farmers turned up for their fuel, the kid was able to work out which tank held the unleaded/ alcohol blend and to fill the other two orders correctly.

How?

224. The Wire's Diameter

At technical school we study the construction of lathes and machines. We learn how to use instruments and how not to be stumped by difficult situations.

My foreman-teacher handed me some wire and asked: "How do you measure a wire's diameter?"

"With a micrometer gauge."

"And if you don't have one?"

On thinking it over, I had an answer. What was it?

225. The Bottle's Volume

If a bottle, partly filled with liquid, has a round, square or rectangular bottom which is flat (see figure above), can you find its volume using only a rule? You may not add or pour out liquid.

226. The Ship and the Seaplane

A diesel ship leaves on a long voyage. When it is 180 miles from shore, a seaplane, whose speed is 10 times that of the ship, is sent to deliver mail.

How far from shore does the seaplane catch up with the ship?

227. The Ships and the Lifebuoy

Two diesel ships leave a pier simultaneously, the *Neptune* downstream and the *Poseidon* upstream, with the same motive force.

As they leave, a lifebuoy falls off the *Neptune* and floats downstream. An hour later both ships are ordered to reverse course.

Will the *Neptune's* crew be able to pick up the buoy before the ships meet?

228. Equation to Solve in Your Head

$$6,751x + 3,249y = 26,751$$
$$3,249x + 6,751y = 23,249$$

229. Three Men in the Street

Three men met on the street—Mr. Black, Mr. Gray and Mr. White.

"Do you know," asked Mr. Black, "that between us we are wearing black, gray and white? Yet not one of us is wearing the color of his name?"

"Why, that's right," said the man in white.

Can you say who was wearing which color?

230. The Square Field

A farmer had a square field with 4 equally spaced oaks in it standing in a row from the center to the middle of one side, as shown in the figure below.

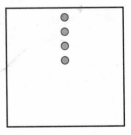

In his will, he left the square field to his 4 sons "to be divided up into 4 identical parts, each with its oak."

How did the sons divide up the land?

231. Kings and Queens

Three playing cards have been removed from an ordinary pack of cards and placed face down in a horizontal row. To the right of a King there are one or two Queens. To the left of a Queen there are one or two Queens. To the left of a Heart there are one or two Spades. To the right of a Spade there are one or two Spades.

What are the three cards?

232. Even Tread

I keep one spare tire in my car. Last year, I drove 10,000 miles in my car, and rotated the tires at intervals so that, by the end of the year, each of the five tires had been used for the same number of miles. For how many miles was each tire used?

233. Round and Round

Imagine Wheel A, with diameter x, rolling around fixed Wheel B, with diameter $2x$. How many revolutions about its own axis will Wheel A make in rolling once around Wheel B?

234. Choose a Glass

Some detectives were investigating a case of poisoning at a hotel. They had lined up a number of partly filled glasses on a table in the hotel lounge, knowing that only one glass contained poison. They wanted to identify which one before testing it for fingerprints. The problem was that, if the police laboratory were asked to test the liquid in each glass, it would take too long. So the inspector in charge contacted a statistician at the local college to see if there was a quicker way. He came over to the hotel, counted the glasses, smiled, and said:

"Choose any glass, Inspector, and we'll test it first."

The inspector was worried that this would mean the waste of one test but the statistician denied this.

Later that evening, the inspector was telling his wife about the incident.

"How many glasses were there to start with?" she asked.

"I don't remember exactly—somewhere between 100 and 200 I think," replied the inspector.

Can you work out the exact number of glasses? (Assume that any group of glasses can be tested simultaneously by taking a small sample of liquid from each, mixing the samples and making a single test on the mixture.)

235. The Square Window

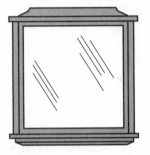

Shown above is a store window that measures 7 feet high by 7 feet wide. The store decorator wants to paint half the window blue and still have a square, clear section of window that measures 7 feet high by 7 feet wide. How would he do this?

236. Dominoes

If you were challenged to cover the board with 32 dominoes, each 1 × 2 inch domino exactly covering two squares on the board, you could solve it in many ways. The checker board above has only 62 squares (two corner squares have been removed), and you must cover this board with 31 dominoes. Can you find a solution? Or if you can't, can you explain why no one ever will?

237. A Bridge Game

In the game of bridge, all the cards are dealt to 4 players—13 cards to each—who usually play as partners, one pair against the other. That should be all you need to know in order to answer the following two questions about situations that might arise in a game.

FLUSH. You and your partner have been dealt a surprising hand. Together you have all 13 cards of one suit. Is this event more or less likely than one in which you and your partner together have no cards in one of the suits?

PAPER PERFECT. Every few years a newspaper story will report that players at a local bridge game were witness to a "perfect deal": that is, each player got all 13 cards of a suit. How many of these deals would you expect to occur anywhere in the world during this decade?

238. Computers

A computer buff called Hacker owns several computers. All but two of them are Apples, all but two of them are Commodores and all but two of them are IBMs. How many computers does our friend Hacker own?

239. Third of the Planet

How far from the center of the Earth would you have to be to see one-third of the planet's surface?

240. Checkers

Take 16 checkers—8 black and 8 white—and arrange them as shown on the overleaf, in a 4 × 4 square with the colors alternating in checkerboard fashion. If you don't have checkers, pennies will do, with the "head" and "tail" sides substituting for the two different colors.

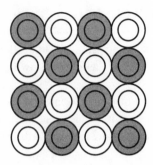

The problem is to rearrange the checkers within a 4 × 4 square so that each column is just one color—either all black or all white. You could easily solve the problem if you were allowed to touch 8 checkers in the square and, with a little figuring, you may find ways to do it by touching only 6. The problem can be solved, however, by touching only 2 checkers in the whole array. How would you do it?

241. The Handicap Race

Mel and Sid race each other in a 100-yard dash. Mel wins by 10 yards. They decide to race again, but this time, to make things fairer, Mel begins 10 yards behind the starting line. Assuming they both run with the same constant speed as before, who wins this time, Mel or Sid? Or is it a draw?

242. . . . 9, 10

Where would you place 9 and 10 to keep the sequence going?

$$1 \quad 2 \qquad\qquad 6$$
$$3 \quad 4 \quad 5 \qquad 7 \quad 8$$

243. The South Pole

Base to explorer at the South Pole: "What's the temperature?"

"Minus 40 degrees," said the explorer.

"Is that Centigrade or Fahrenheit?" asked base.

"Put down Fahrenheit," said the explorer. "I don't expect it will matter."

Why did he say that?

244. Guinness or Stout

Two strangers enter a pub. The bartender asks them what they would like.

First man says, "I'll have a bottle of stout," and puts 50¢ down on the counter.

Bartender: "Guinness at 50¢, or Jubilee at 45¢?"

First man: "Jubilee."

Second man says, "I'll have a bottle of stout," and puts 50¢ on the counter. Without asking him, the bartender gives him Guinness.

How did the bartender know what the second man was drinking?

245. Bonus Payments

A company offered its 350 employees a bonus of $10 to each male and $8.15 to each female. All the females accepted, but a certain percentage of the males refused to accept. The total bonus paid was not dependent upon the number of men employed. What was the total amount paid to the women?

246. The Tiled Floor

A $2\frac{1}{2}$-inch-square card is thrown at random on to a tiled floor (see figure below). What are the odds against its falling and not touching a line? You should assume that the pattern repeats over a large area.

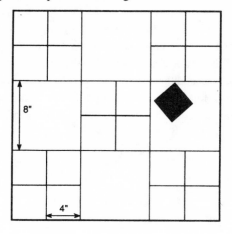

247. Shuffled Deck

Prove that if the top 26 cards of an ordinary shuffled deck contain more red cards than there are black cards in the bottom 26, then there are in the deck at least 3 consecutive cards of the same color.

248. A Peculiar Number

If a certain number is reduced by 7 and the remainder is multiplied by 7, the result is the same as when the number is reduced by 11 and the remainder is multiplied by 11. Find the number.

249. Antifreeze

A 21-quart capacity car radiator is filled with an 18 percent alcohol solution. How many quarts must be drained and then replaced by a 90 percent alcohol solution for the resulting solution to contain 42 percent alcohol?

250. Tree Leaves

If there are more trees than there are leaves on any one tree, then there exist at least two trees with the same number of leaves. True or false?

251. The Will

Daniel Greene was killed in a car crash while on his way to the maternity hospital where his wife, Sheila, was about to give birth. He had recently made a new will, in which he stated that, should the baby be a boy, his estate was to be divided two-thirds to his son and one-third to Sheila; if the baby were a girl, then she was to receive a quarter of the estate and Sheila the other three-quarters.

In the event Sheila gave birth to twins—a boy and a girl. How best should Daniel's estate be divided so as to carry out his intention?

252. Watered-down Wine

Imagine you have two large pitchers. One contains a gallon of water and the other a gallon of wine. One pint of wine is removed from the wine pitcher, poured into the water pitcher and mixed thoroughly. Then a pint of the mixture from the water pitcher is removed and poured into the wine pitcher.

Is there now more or less water in the wine pitcher than there is wine in the water pitcher?

253. A Logic Riddle

In olden days, the student of logic was given this problem: If half of 5 were 3, what would one-third of 10 be?

254. A Matter of Health

If 70 percent of the population have defective eyesight, 75 percent are hard of hearing, 80 percent have sinus trouble and 85 percent suffer from allergies, what percentage (at a minimum) suffer from all four ailments?

255. Streetcars

A man is walking down a street along which runs a streetcar line. He notices that, for every 40 streetcars which pass him traveling in the same direction as he, 60 pass in the opposite direction. If the man's walking speed is 3 miles per hour, what is the average speed of the streetcars?

256. Passing trains

A man and a woman are walking along a railway track. A train passes the man in 10 seconds. Twenty minutes later, it reaches the woman. It passes her in 9 seconds. How long after the train leaves the woman will the man and woman meet if all speeds are constant?

257. The Fly and the Record

A fly is walking around the groove of a 33 r.p.m. record. The record is lying flat on the floor, and when looked at from above, the fly appears to be traveling clockwise. If it carries on in this way, will it eventually arrive at the edge of the record or the center?

258. Move One Coin

Ten coins are arranged as shown above. Can you move just one coin to another position so that, when added up either horizontally or vertically, two rows of 6 coins each will be formed? (It's best to try to find the solution using real coins.)

259. The Unbalanced Coin

You have a coin that you have reason to suspect is unbalanced; that is, it is biased towards heads or tails, and a long series of tosses won't come out 50-50. How can this coin be used to generate a series of random binary digits—ones and zeros?

260. Bicycle Experiment

A rope is tied to a bicycle pedal that is stationary at the bottom of its arc (see above). If someone pulls back on the rope while another person holds the seat lightly to keep the bicycle balanced, will the bicycle move forward, backward, or not at all?

261. A Piece of String

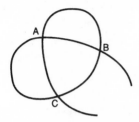

A piece of string, lying on the floor in the pattern shown above, is too far away for an observer to see how it crosses itself at points A, B and C. What is the probability that the string produces a knot if the ends are pulled apart?

262. The Island and the Trees

The figure below shows a deep circular lake, 300 yards in diameter, with a small island at the center. The two black spots are trees.

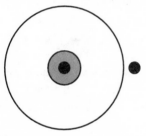

A girl who cannot swim has a rope a few yards longer than 300 yards. How does she use it as a means of getting to the island?

263. A Boy, a Girl and a Dog

A boy, a girl and a dog start at the same spot on a straight road. The boy walks forward at 4 miles per hour; the girl walks forward at 3 miles per hour. Meanwhile, the dog trots back and forth between them at 10 miles per hour. Assuming that each reversal of direction of the dog is instantaneous, where is it and which way is it facing after one hour?

264. Boxes and Balls

Four blindfolded girls were each given an identical box, containing different colored balls. One contained 3 black balls; one contained 2 black balls and 1 white ball; one contained 1 black ball and 2 white balls; and the fourth contained 3 white balls. Each box had a label on it reading "Three Black" or "Two Black, One White" or "One Black, Two White" or "Three White." The girls were told that none of the four labels correctly described the contents of the box to which it was attached. Each girl was told, in turn, to draw 2 balls from her box, at which point her blindfold was removed so that she could see the 2 balls in her hand and the label on the box assigned to her. She was then given the task of trying to guess the color of the ball remaining in her box.

As each girl drew balls from her box, their colors were announced for all the girls to hear; but the girls could not see the labels on any box other than their own.

The first girl, having drawn 2 black balls, announced: "I know the color of the third ball." The second girl drew one white and one black ball, and similarly stated: "I know the color of the third ball." The third girl withdrew two white balls, looked at her label, and said: "I can't tell what the color of the third ball is." Finally, the fourth girl declared: "I don't need to remove my blindfold or any balls from my box, and yet I know the color of all three. What's more, I know the color of the third ball in each of your boxes as well."

The first three girls were amazed by the fourth girl's assertion and immediately challenged her. She proceeded to identify the contents of each box correctly. How?

265. Two Trains

If it takes twice as long for a passenger train to pass a freight train after it first overtakes it as it takes for the two trains to pass each other when going in opposite directions, how many times faster than the freight train is the passenger train traveling?

266. Missing Elevation

From the front elevation and the plan opposite, can you find the side elevation and describe the object?

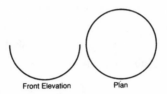

Front Elevation   Plan

267. Avoiding the Train

A man was walking down a railway track when he saw an express train speeding towards him. To avoid it he jumped off the track, but before he jumped he ran 10 feet towards the train. Why?

268. Bowl and Pan

My mother has a bowl that holds a little more than a pint, and a flat rectangular, straight-sided pan that holds exactly a pint. The figure below shows its proportions.

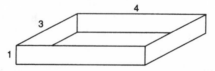

She wants to put exactly one-third of a pint of water into the bowl, but has no other means of measuring anything. She has a supply of water and an ordinary kitchen table with an exactly level surface. How does she do it?

269. Jasmin's Age

When Jasmin went to the polling station to vote, the clerk asked her age.

She told him, "Eighteen."

He looked at her, surprised, and said, "Are you sure that's right?"

Jasmin laughed and replied, "No, I gave myself the benefit of a year less than a quarter of my real age."

The clerk couldn't calculate how old she actually was, though he did allow her to register her vote. What was Jasmin's age?

270. A Ball of Wire

A wire of $\frac{1}{100}$ of an inch diameter is tightly wound into a ball with a diameter of 24 inches. It is assumed that the wire is bound so solidly that there is no air gap in the ball. What is the length of the wire?

271. Ferry Boats

Two ferry boats start moving at the same instant from opposite sides of the Hudson River, one boat going from New York to Jersey City, and the other from Jersey City to New York. One boat is faster than the other, so they meet at a point 720 yards from the nearest shore.

After arriving at their destinations, each boat remains for 10 minutes to change passengers before starting on the return trip. The boats meet again at a point 400 yards from the other shore.

What is the exact width of the river?

272. John and the Chicken

John was attempting to steal a chicken. When he first saw the bird, he was standing 250 yards due south of it. Both began running at the same time and ran with uniform speeds. The chicken ran due east. Instead of running northeast on a straight line, John ran so that at every instant he was running directly towards the chicken.

Assuming that John ran $1\frac{1}{3}$ times faster than the chicken, how far did the chicken run before it was caught?

273. The Prisoner's Choice

A prisoner was about to be executed but was promised his freedom if he drew a silver ball from one of two identical urns. He was allowed to distribute 50 silver and 50 gold balls between the two urns any way he liked. The urns were then going to be shuffled around out of his sight and he was to pick one urn and draw one ball at random from that urn.

How did the prisoner maximize his chances of success? If he had put equal numbers of silver and gold balls into one of the urns, the other urn would also contain equal numbers of silver and gold balls, and thus the probability of his drawing a silver would have been 1 in 2. Can you improve these chances, and if so, how?

274. Counters in a Cup

I hesitated over whether to include this puzzle, as it straddles the fence between a trick question and a legitimate puzzle. However, the solution is especially neat.

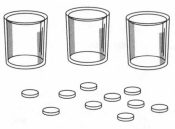

How can 10 counters be distributed between 3 cups (see above) so that each cup contains an odd number of counters?

275. Speed of Ant

A subway train is approaching Grand Central Station at 114 inches per second. A passenger in one car is walking forward at 36 inches per second relative to the seats. He is eating a footlong hot dog, which is entering his mouth at the rate of 2 inches per second. An ant on the hot dog is running away from the passenger's mouth at 1 inch per second. How fast is the ant approaching Grand Central?

276. Wayne and Shirley

Wayne and Shirley have agreed that they would like to have a family of 4 children, but they would prefer not to have them all the same sex. Is it more likely that they will have 3 of one sex and 1 of the other or 2 of each? (Assume that each birth has an equal chance of being a boy or a girl, which, in the real world, is not statistically quite the case.)

277. Word Series

What is the next word in the following series: *aid, nature, world, estate, column, sense* . . . ?

Is it (a) *water,* (b) *music,* (c) *welcome,* or (d) *heaven*?

278. Shooting Match

Two sharpshooters, Bill and Ben, had a contest to see which of them was the better shot. In their first session, each fired 50 rounds and hit the target 25 times. Later, they had a second session, and this time Bill hit the target 3 times in 34 shots, and Ben missed 25 shots in a row before giving up. Since Bill's record in the second session was better than Ben's, Bill argued that his record for the two sessions combined was better than Ben's. Was he right?

279. Lethargic Llamas

A zoo keeper houses 9 llamas all together in a large square cage.

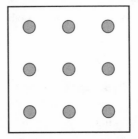

Strangely, all these llamas are very lethargic and remain in the positions shown in the figure above at all times. Can you give each its own private cage by building just 2 more square enclosures?

280. Torpid Tapirs

The same zoo also has 10 torpid tapirs which are constantly positioned in a circular pen as shown below.

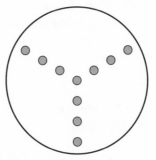

Give each tapir its own private compartment by drawing only 3 additional circular enclosures.

281. Caroline and the Lake

Caroline decided to leave her husband and went to stay in a women's refuge near Lake Round, a large artificial body of water so named because of its precisely circular shape. However, her husband followed her there. To escape him, she took a rowing boat and rowed to the center of the lake, marked by a buoy. Her husband was furious, but, knowing she would have to come ashore eventually, decided to wait for her on the edge of the lake. He assumed that, because he could run 4 times as fast as Caroline could row, it would be a simple matter for him to catch her when the boat touched the edge of the lake.

Caroline thought about her predicament. She knew that once she was on dry land she could run faster than her husband, so she devised a plan that got her to some point on the shore before he could get there. What was her strategy? You can assume that at all times she knew her exact position on the lake.

282. Wooden Block

The wooden block shown above has been cut into two pieces and then reassembled. The pattern on the two hidden sides is the same as that on the two visible sides. How was this done?

283. Word Affinity

All but one of the following have something in common. What is it, and which is the odd one out?

SOCK   SICK   BRICK   BUS   POCKET

FIELD   LINE   TIGHT   FORCE   LIFT

284. Gun Problem

I have a gun I use for clay-pigeon shooting. This gun is 1.4 yards long. One day I wanted to take the train to join a shooting party, but the ticket clerk told me that it was against the regulations for a passenger to take a firearm into the car, and it couldn't be put in the baggage area either because the baggage handler wasn't allowed to take any articles whose greatest dimension exceeded 1 yard. What did I do to ensure that both I and my gun were allowed on the train?

285. Counter Colors

A bag contains one counter, which may be either black or white. A second counter, which is definitely white, is put into the bag. The bag is shaken and one counter taken out, which proves to be white. What is the probability of the next counter coming out of the bag also being white?

286. Bank Account

A bank customer had $100 in his account. He then made 6 withdrawals, totaling $100. He kept a record of these withdrawals, and the balance remaining in the account, as follows:

Withdrawals	Balance left
$50	$50
25	25
10	15
8	7
5	2
2	0
$100	$99

When he added up the columns as above, he assumed that he must owe $1 to the bank. Was he right?

287. Hat in the River

A man wearing a straw hat was fishing from a rowing boat in a river that flowed at a speed of 3 miles per hour. The boat drifted down the river at the same rate.

Just as the man started to row upstream, the wind blew his hat off his head and into the water beside the boat. However, he didn't notice that his hat was gone until he had rowed 5 miles upstream, at which point he immediately started rowing back downstream to retrieve his hat.

The man's rowing speed in still water is a constant 5 miles per hour. However, when rowing upstream his speed relative to the shore would be only 2 miles per hour, given the rate of flow of the river. Rowing downstream, his speed relative to the shore would be 8 miles per hour for the same reason.

If the man lost his hat at 2 o'clock in the afternoon, what was the time when he retrieved it?

288. Tossing Pennies

Jill offered Jack the following bet: she said she would toss 3 pennies in the air, and if they fell all heads or all tails she would give him $1. If they fell any other way, he had to give her 50 cents. Should Jack accept?

289. The Kings

Six playing cards are lying face down on the table. Two, and only two, of them are Kings, but you don't know which. You pick two cards at random and turn them face up. Which is more likely:

a) That there will be at least one King among the two cards; or

b) That there will be no King among the two cards?

290. Traffic Lights

The main street in our town has a linked traffic system consisting of 6 consecutive sections, each a multiple of $\frac{1}{8}$ of a mile in length, and each terminating in a traffic light. These traffic lights have a 26-second cycle, which can be considered as 13 seconds on red and 13 seconds on green. The lights are synchronized so that a vehicle traveling at 30 miles per hour will pass each light at the same point in its cycle.

My brother, Albert, has studied the system and reckons he can drive faster than 30 miles per hour and still get through the entire system without crossing a red light. Recently, he set up an experiment with the help of two friends, Robert and Hubert. All three entered the first section simultaneously, Albert traveling at 30, Hubert at 50 and Robert at 75 miles per hour, with the first traffic light turning green 3 seconds later.

Robert got through the whole system in less than 2 minutes without being stopped. However, he thought he had been lucky, as he arrived at the last light just as it changed to red. Hubert ran out of petrol after the third light, and in any case would have been stopped at the second light had he not lost 10 seconds due to a delay in the second section.

What were the lengths of each of the 6 sections?

291. The Feast Day

In a remote village in the Himalayas, a Feast Day is declared whenever the bells of the temple and the monastery ring at exactly the same time. The temple bell rings at regular intervals of a whole number of minutes. The monastery bell also rings at regular intervals, but of a different whole number of minutes. Today, the bells are due to ring together at 12 noon.

Between Feast Days, the bells of the temple and monastery ring alternately, and although they only coincide on Feast Days, they do occur as little as a minute apart on some of the other days.

The last time the bells coincided was at 12 noon a prime number of days ago. How many days ago was that?

292. The Clock-mender

I have two clocks which, when fully wound, will run for nearly 8 days before stopping. Both of these clocks were keeping different times, and each was wrong by an exact number of minutes per day, though less than 1 hour in each case.

I took my clocks to be fixed. The local clock-mender works from 9:30 A.M. to 5:00 P.M., Mondays to Fridays. He immediately wound both clocks fully and set them to the right time—a whole number of minutes after the hour—then put them on a shelf for observation.

The following Monday, as he went to take down the clocks to start work on them, they both started to strike 8 o'clock simultaneously. This was some hours and minutes past the correct time. What day and exact time was it when the clock-mender set them originally?

293. The Typewriter

My typewriter used to have a standard keyboard, with the letters arranged as follows:

Row 1: Q W E R T Y U I O P

Row 2: A S D F G H J K L

Row 3: Z X C V B N M.

I lent the machine to a friend, and when it came back I found that the positions of the letters had been altered into what he claimed was a more efficient layout. None of the letters was in its original row, though the numbers of letters in each row remained the same.

I tested the new layout by typing various words connected with my business—I run a liquor department in a supermarket in London. The numbers in parentheses below show the number of rows I had to use to produce each word:

BEER (1)
STOUT (1)
SHERRY (2)
WHISKEY (3)
HOCK (2)
LAGER (2)
VODKA (2)
CAMPARI (2)
CIDER (3)
SQUASH (2, but would have been 1 but for a single letter)
FLAGON (2)
MUZZY (2).

The next word I tried was CHAMBERTIN. Which rows, in order, did I use?

294. The Bridge

A-town and B-town are two villages connected by a bridge spanning a river. At the end of a war, the occupying forces installed a sentry in the middle of the bridge to prevent the inhabitants of A-town and B-town from visiting each other. All means of transport having been requisitioned, the only access from village to village is by foot over the bridge, which would take 10 minutes. The sentry is under strict orders to come out of his bunker every 5 minutes, and send anyone trying to cross back to his own village, if necessary by force of arms. Michael in A-town is desperate to visit his girlfriend in B-town. Is there a way?

295. The Cookie Jar

An old nursery rhyme starts: "Who stole the cookie from the cookie jar . . ."
Let us find out from the following statements, of which only one is true:

ANN: Harry stole the cookie from the cookie jar.

HARRY: Fred stole the cookie from the cookie jar.

LISA: Who me?—can't be.

FRED: Harry is lying when he says that I stole the cookie.

296. Crossing the Desert

A small airplane carrying 3 men has to make an emergency landing in the middle of the desert. The men decide that their best chance for survival is for each of them to set out across the desert in a different direction, in the hope that one of them will be able to reach civilization and get help for the others. Their supplies include 5 full bottles of water, 5 half-full, and 5 empty bottles.

Since water-carrying capacity is important should a man reach an oasis, they wish to divide both the water supply and the number of bottles equally among themselves. How can they achieve this?

297. Panama Canal

A ship entered the Panama Canal at its west end, passed through the canal, and left at its east end. However, immediately after it left the canal, it entered the Pacific Ocean. If the ship did not double back or sail backwards, how could this be?

298. The Short Cut

John was trying to take a short cut through a very narrow tunnel when he heard the whistle of an approaching train behind him. Having reached three-eighths of the length of the tunnel, he could have turned back and cleared the entrance of the tunnel running at 10 miles per hour just as the train entered. Alternatively, if he kept running forward, the train would reach him the moment he would jump clear of the tracks. At what speed was the train moving?

299. Red, White and Blue

This is a famous paradox which has caused a great deal of argument and disbelief from many who cannot accept the correct answer.

Four balls are placed in a hat. One is white, one is blue and the other two are red. The bag is shaken and someone draws two balls from the hat. He looks at the two balls and announces that at least one of them is red.

What are the chances that the other ball he has drawn out is also red?

300. Common Factor

What do the following words have in common?

DEFT   FIRST   CALMNESS   CANOPY
LAUGHING   STUPID   HIJACK.

301. Word Groups

Which word from Group 2 belongs with those in Group 1?

Group 1:   BAG   STORM   BANK   BAR
Group 2:   MOON   FLOOR   STORE   DUNE

302. Two Wins

Bill is a keen chess player and often plays against his parents. He wins and loses against both parents, but his mother is a better player than his father.

His parents offer to double his allowance if he can win two games in a row out of three, with his parents alternating as opponents. Which parent should he play first to maximize his chances of winning two in a row?

303. Find X

Solve the following equation for *x*:

$$\sqrt{x+\sqrt{x+\sqrt{x}\ \ldots}} = 2$$

304. Pocketful of Coins

Freddy has 10 pockets and 44 coins. He wants to distribute the change among his pockets so that each pocket contains a different number of coins. Can he do this?

305. Six-gallon Hat

You bought a ten-gallon hat as a souvenir of a visit to Texas; only when you got home did you discover that the label states it to be only a six-gallon hat. By now, you were skeptical that it was even that big, and you decided to test it by trying to fill it with 6 gallons of water. The only containers you had to hand were those below. Using them, how were you able to pour 6 gallons into the hat?

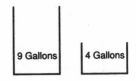

306. A Flock of Geese

Two brothers, Guy and George, inherited a flock of geese. They sold the entire flock, receiving for each goose the same number of dollars as there were geese in the flock. The money was given to them in $10 bills except the odd amount, less than $10, which was paid in change. They divided the bills by dealing them out alternately, though Guy complained that this was not fair because George received both the first and last bills, thus getting $10 more. To even things up, George gave Guy all the change, but Guy argued he was still worse off. George agreed to give Guy a check for the difference. What was the value of the check?

307. Three Points on a Hemisphere

Three points are selected at random on a sphere's surface. What is the probability that they all lie in the same hemisphere? Assume that the great circle, bordering a hemisphere, is part of the hemisphere.

308. Deal a Bridge Hand

A man had dealt about half the cards for a bridge game when he was interrupted by a telephone call. When he returned, no one could remember who had been dealt the last card. Without knowing the number of cards in any of the 4 partly dealt hands, or the number of cards left in the undealt part of the pack, how could the deal be completed so that everyone received the cards they would have received had the deal not been interrupted?

309. The Fifty-dollar Bill

The owner of the local bank found a $50 bill lying in the gutter; he picked it up and made a note of its serial number. Later that day his wife mentioned that they owed the butcher $50, so the banker used the bill he'd found to settle up with the butcher. The butcher used it to pay a farmer; the farmer in turn used it to pay his feedstock supplier; and the feedstock supplier used it to pay his laundry bill. The laundryman used it to pay off his $50 overdraft at the local bank. The banker recognized the bill as being the one he had found in the gutter, but also noticed, on closer examination, that it was a fake. By now, it had been used to settle $250 worth of debts.

What was lost as a result of this series of transactions, and by whom?

310. The Bicycle Race

Two cyclists are racing around a circular track. Pierre can ride once around the track in 6 minutes, Louis takes 4 minutes. How many minutes will it take for Louis to lap Pierre?

311. The North Pole

A man goes to the North Pole. The points of the compass are, of course:

```
        N
    W       E
        S
```

He reaches the Pole and, having passed over it, turns about to look North. East is now on his left-hand side, and West on his right-hand side, so the points of the compass appear to be:

```
        N
    E       W
        S
```

Is this correct? If not, what is the explanation?

312. Card Games

Jack and Jill are playing cards for a stake of $1 a game. At the end of the evening, Jack has won 3 games and Jill has won $3.

How many games did they play?

313. Long-playing Record

The diameter of a vinyl long-playing record is 12 inches. The unused part in the center has a diameter of 4 inches and there is a smooth outer edge 1 inch wide around the recording. If there are 91 turns of the groove to the inch, how far does the needle move during the actual playing of one side of the record?

314. Which Games?

I have three friends. Two play football, two play tennis and two play golf. The one who does not play golf does not play tennis, and the one who does not play tennis does not play football. Which games does each friend play?

315. What Day Is It?

When the day after tomorrow will be yesterday, today will be as far from Sunday as today was from Sunday when the day before yesterday is tomorrow. What day is it today?

316. Cash Bags

A man went into a bank with exactly $1,000 in silver dollars. He gave them to a cashier and asked the cashier to put the money into 10 bags in such a way that if he later needed any amount of dollars up to $1,000, he could lay his hands on that amount without needing to open any of the bags. How did the cashier achieve this?

317. Garage Space

A haulage contractor did not have room in his garage for 8 of his trucks. He therefore increased the size of his garage by 50 percent, which gave him room for 8 more trucks than he owned altogether. How many trucks did he own?

318. Bag of Chocolates

Three girls agreed to share out a bag of chocolates in proportion to their ages. The sum of their ages was $17\frac{1}{2}$ years, and the bag contained 770 chocolates. For every 4 chocolates Joan took, Jane took 3; for every 6 Joan took, Jill took 7. How many chocolates did each girl take, and what are their respective ages?

319. Lost

I am traveling in a strange country. I have no map. I come to a crossroads where the signpost has been knocked down. How can I find my way without asking anyone for directions?

320. Problem Age

The day before yesterday, Peter was 17. Next year he will be 20. How do you explain this?

6. Chess Problems (Some Are, Some Aren't)

Conventional chess problems are outside the scope of this book for several reasons. Obviously, they depend on specialized knowledge and interest; they enjoy their own vast literature and, unlike the brainteasers in this collection, chess allows for an infinite number of variations based on the same essential premise. No matter how often one tackles "Mate in two moves" puzzles, the experience does not make answering the next "Mate in two moves" a foregone conclusion.

On first reading, the following puzzles seem as though they would be perfectly at home in the chess column of your daily newspaper, and, in fact, they *do* require an understanding of the rudiments of chess. (Actually, one of them hardly requires even that, and another *definitely* does not.) But—with one exception—they are not chess problems in the conventional sense, any more than a puzzle which requires a basic knowledge of, say, the solar system would be thought of as a puzzle for astronomers only. Put another way, the puzzles contained in this chapter are about situations occurring on the chessboard, but the leap of imagination needed to solve them is such that a pedestrian player might succeed where a chess grandmaster would fail.

321. Mate in One

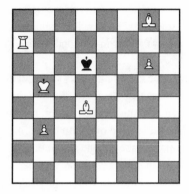

How did White mate in one move?

322. Endgame

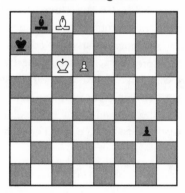

It is well known that Bishops of opposite colors often lead to an endgame draw. The endgame shown above is complicated, however, by advanced Pawns on both sides. Can White (to move) force a win against Black's best defense?

323. Smart Kid

While I have long been an ardent chess player, my 12-year-old daughter barely knows the moves. (The reader does not need to know either, for this puzzle!) Recently two of my friends, who are chess experts, came to dinner. After dinner, I played a game against each of them, losing both games even though my guests had given me a pawn advantage and the opening move. My daughter had wandered into the room midway through the first game, and she watched, glumly, as her dad got trounced. "Daddy, I'm ashamed of you," she said, as I started putting the pieces away after my second defeat. "I could do better than that."

How cute, my guests and I thought, until she became so insistent she be given the chance to play them that we got to thinking that maybe she needed to be taught a lesson. When she announced: "I don't want any advantage—I'll play one game with white pieces and one with black (meaning she would have the first move in one game and the second in the other). In fact, I'll give *them* an advantage by playing both games at once!" my friends were ready to wipe the chessboard with her. And, I have to confess, I was quite looking forward to watching them do it.

Did they?

324. On the Move

This is not, in any sense, a chess problem, but it is included here because the setting is a chessboard or, more particularly, one column of a chessboard. Picture two players sitting opposite each other in the traditional positions of White and Black. They each have one Rook, one Knight and one Pawn lined up along one of the rows between them as shown below, so that there is one empty square (4) between the opposing "armies." As there are 8 squares in a row on a chessboard, one of the end squares must be ignored.

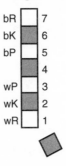

As you can see, for the purpose of this puzzle, the squares are labeled 1 to 7 and the pieces w (white) and b (black) R (Rook), K (Knight) and P (Pawn).

The object is to move the white pieces to the squares occupied by the black pieces and vice versa, observing the following rules:

1. Pieces can only move forward.
2. One square at a time.
3. Except that they can jump over *one* piece of the opposite color, thus moving 2 squares, provided the square on the other side of the piece being jumped is empty.

An intriguing aspect of this problem is that, even if demonstrated, at speed, to someone who hasn't been able to work out the solution, it will take several attempts before the exercise can be repeated.

325. Checkmate

Just as I entered my chess club, I saw Michael declare checkmate. He and his opponent shook hands and proceeded to the bar. I looked at the board (below) and found that White had indeed checked Black, but how had he done it?

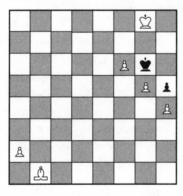

326. White to Move
BLACK

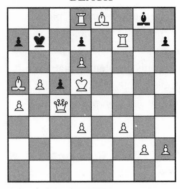

WHITE

White to move, mates in one.

327. Charles XII

Sam Loyd's most widely known chess problem is strictly for chess enthusiasts, but I give it here, without apology, because of its exceptionally intriguing nature and elegant construction, which non-chess players may appreciate even if the solution is beyond them. Indeed, the solution is so intricate that it is not, customarily, given in the many reprintings the puzzle has enjoyed. The chess-lover is, therefore, left to ponder it at his or her leisure.

The story is told of Charles XII of Sweden, who was besieged by the Turks in 1713 at his camp in Bender. The King often passed the time by playing chess with one of his ministers. On one occasion, the game reached the situation depicted on the overleaf.

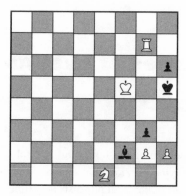

King Charles, playing White, announced a checkmate in three moves. At that instant, a bullet shattered the white Knight. The King studied the board again, smiled, and said he did not need the Knight because he still had a mate in four moves. No sooner had he said this than a second bullet removed his Pawn at King's Rook 2. Charles reconsidered his position carefully and announced mate in five.

The story has a twist. Years later, a German chess expert pointed out that if the first bullet had destroyed the white Rook instead of the Knight, Charles still would have had a mate in six.

7. Puzzles of Everyday Life

Puzzle-solving has been a popular hobby for several centuries. Challenging and intellectually satisfying puzzles are usually the product of some considerable ingenuity. The work of the great puzzlists is as timeless as the great works of literature, and it is not meant as a criticism to concede that many of the best problems are somewhat artificial confections, depending on fairly improbable sets of circumstances and a certain suspension of disbelief on the part of the reader.

At the same time, the world around us is a good provider of intriguing puzzles in its own right, and this chapter is devoted to some curiosities and conundra of everyday life. You will probably be familiar with at least some of the following, though the more than usually fulsome "solutions" which I offer may still provide the odd surprise. Most of the rest are quite easy, though some, I suspect, will only seem so in retrospect, after a little agonizing to find the solution or even having to look it up at the back of the book. And that, arguably, is the hallmark of the truly classic brainteaser, whether man-made or existing in the real world: an infuriating blend of the obvious and the elusive.

328. Railways
The rails in the American and British railway systems have a length of 60 feet, while most of the railways in continental Europe use sections of 30 meters (98 feet 5 inches).

You will notice that railway tracks have a small gap between adjoining sections. Why?

329. Rice and Salt
In some countries, it is the custom to put rice in salt dispensers. Why?

330. Coal and Lime

In some European countries, if coal is transported in open railway trucks the top is sprayed with a solution of lime. Why?

331. Sand on the Beach

Walk along a beach at low tide when the sand is firm and wet. At each step the sand immediately around your foot dries out and turns white. Why? The popular answer, that your weight "squeezes the water out," is incorrect: sand doesn't behave like a sponge. So what does cause the whitening?

332. Bridge Clearance

A trucker wants to drive under a bridge but finds that his rig is an inch higher than the bridge's clearance. The frustrated driver pulls to the side of the road and is checking maps to find his shortest alternative route when a small child comes up to him and says, "Hey, Mister, for five bucks I'll tell you how to get your truck through." Her suggestion worked. What was it?

333. A Flat Tire

A man whose car suddenly blows a tire pulls up at the side of the road. He jacks up the car, removes the hubcap and 4 lug nuts, places the nuts in the hubcap, and removes the flat. As he is lifting the spare wheel out of the boot, however, he kicks over the hubcap and all the nuts fall into a drain. The next town is several miles away. Of course, he could walk or hitchhike there to buy more lug nuts; but can you think of anything better?

334. The Biology Exam

Scot Morris, in *The Next Book of Omni Games,* relates the story of a biology final exam.

The professor warned the students that everyone had to stop writing when the 3 o'clock bell rang, but one young man ignored him and continued writing for a few seconds more. When he dropped his exam booklet on top of the others, the professor picked it up and handed it back. "You were told to stop writing when the bell rang. Since you disobeyed, you fail the course."

The student demanded indignantly, "Do you know who I am?"
"No, and furthermore, I don't care," the teacher replied.
What did the student do to avoid failing the test?

335. The Manhole

Why are manhole covers circular rather than square?

336. Wet Roof

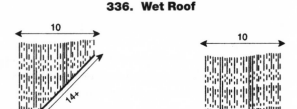

When rain falls straight down onto a roof with a 45° tilt (as illustrated above), there is less rain per unit area than if the roof were level (above right). It would therefore seem that rain falling vertically on level ground would give it more of a wetting than the same rain would if it fell at an angle because of the wind. Why is this not the case?

8. Prove That

This chapter is not for the faint-hearted. It mines what I consider to be a particularly rich vein of intellectual stimulation for the more ambitious puzzlist, one which, surprisingly, the literature of recreational mathematics has tended to ignore in the past.

The history of mathematics spans several millennia; at each turn, in each of its various disciplines, that long road is marked with the discovery of the basic principles, laws, truths—call them what you will—that govern the science we call mathematics. Nowadays, these principles are axiomatic; we take for granted that the angles of a triangle always total 180°, just as we take for granted the existence of gravity or tomorrow's rising of the sun.

But these principles were not always taken for granted. Indeed, in each case, there was a time before the principle was even suspected, and an innovative mathematician who was the first person to recognize and articulate it. In the arena of scientific discovery, there are as many different routes to the truth as there are truths, but they have certain characteristics in common: with a hypothesis as a starting point, a theorem is developed which then has to be proved by unimpeachable reasoning. This process from *supposition*—perhaps, at the outset, little more than a scientist's hunch—to established *proposition* is known by the term "scientific method."

It may be dangerous to describe all the great laws of science as immutable; Newton's laws of gravity, the supposed inability of light to bend and parallel lines to meet are just a few examples of yesterday's hallowed axioms that have been called into question. But they *do* have at least one characteristic in common: to someone with scientific sensibilities, their proofs have an ethereal beauty that comes from being intellectually satisfying.

Many of these propositions and/or their proofs are very complicated or arcane, beyond the scope of a book that is intended to be recreational. But some can be addressed with the simple tools of elementary algebra and geometry (and some, indeed, with little more than common sense). The challenge I issue in this chapter is, in some cases, to retrace mathematical principles (some of which have become axiomatic), and, in others, to prove with scientific precision what might be called an instinctive truism.

337. Interior Angles

Prove that the three interior angles of any triangle total 180° (in Euclid geometry).

338. Angle in Semicircle

Prove that any angle inscribed in a semicircle is a right angle.

339. Pythagoras

Raymond Smullyan, in his book *5000 B.C. and Other Philosophical Fantasies,* tells how he once introduced his students to the Pythagorean theorem:

"I drew a right triangle on the board with squares on the hypotenuse and legs and said, "Obviously, the square on the hypotenuse has a larger area than either of the other two squares. Now suppose these three squares were made of beaten gold, and you were offered either the one large square or the two small squares. Which would you choose?"

Interestingly enough, about half the class opted for the one large square and half for the two small ones. A lively argument began. Both groups were equally amazed when told that it would make no difference."

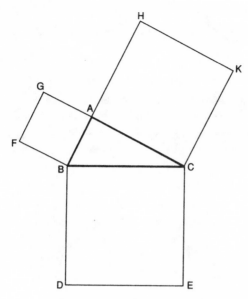

Prove that this is so, i.e. that the sum of the areas of the squares ABFG and ACKH (above) equals square BCED.

340. Lunes

The word "lune" takes its origin from the Latin word *luna,* meaning "moon." In geometry, lunes are plane regions bounded by arcs of different circles (see shaded crescents in the figure below).

Prove that the sum of the areas of two lunes constructed on two sides of a triangle which is inscribed in a semi-circle equals the area of that triangle.

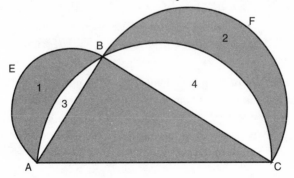

That is, in the above figure, if ABC, AEB and BFC are semicircles, prove that the area of Lune 1 + the area of Lune 2 = the area of the triangle ABC.

341. Diagonals of a Rectangle

Two lines of equal length cross at their centers (see figure below).

Prove that they are diagonals of a rectangle. Don't be tempted to say that this is self-evident, but provide a geometrical proof.

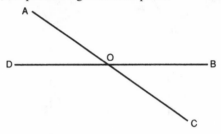

342. Division of Angle Bisector

Prove that the bisector of an angle of a triangle divides the opposite side into segments proportional to the adjacent sides.

343. Inscribed Decagons

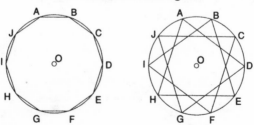

If the circumference of a circle is divided into 10 equal parts (above left), the chords joining consecutive points of division form a regular decagon. The chords joining every third division point form an equilateral star decagon (above right). Show that the difference between the sides of these decagons is equal to the radius of the circle.

344. Three Circles

Draw any three circles.

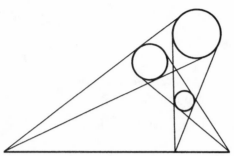

Prove that the intersection points of the common outside tangents to each of the three circles (see figure above) all lie on a straight line.

345. A Thousand Points

The figure below shows some of an infinite number of non-touching points lying inside a closed curve.

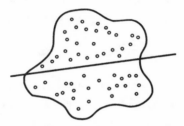

Assuming that any even number of points—say, a thousand—are selected at random, prove that it is possible to place a straight line on the plane so that it cuts the curve, misses every point in the set of a thousand, and divides the set so that half the points lie on one side of the line and half on the other.

346. The Hermit

At sunrise on Monday, a hermit began climbing the narrow path to his hut at the top of a mountain. He did not walk at a constant speed but stopped occasionally to eat or rest, reaching his hut shortly before sunset. On Tuesday morning, he descended the same path, starting at sunrise and again walking at varying rates, though generally at a faster pace than his average speed going up.

Prove that there is a spot along the path that the hermit will occupy on each trip at precisely the same time of day.

347. Volume of a Sphere

Assume that you know the formula for:

The volume of a cone to be: $\dfrac{\text{base} \times \text{height}}{3} = \dfrac{r^2.\pi.h}{3}$

and the surface of a sphere to be: $4R^2.\pi$

Prove by reasoning without the use of calculus that the volume of a sphere is: $\dfrac{4.R^3.\pi}{3}$.

348. Five Points in a Square

Take a one inch square and select five points on or in the square at random. Prove that there are at least two points which are not more than $\dfrac{\sqrt{2}}{2}$ inches apart.

349. The Alternative Triangle

An isosceles triangle has two 13-inch sides and a 10-inch base, as shown.

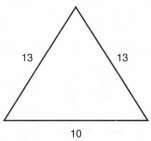

Prove that you can construct another triangle with a different base, having precisely the same area.

350. The Four Integers

Prove with a simple equation that the product of four consecutive integers can never be a perfect square.

351. Three Points on a Sphere

Prove by logical reasoning that any three points chosen at random on a sphere must lie on the same hemisphere.

352. The Three Cities

If city A is 9,000 miles from London and London is 9,000 miles from city B, prove by reasoning that city A must be closer to city B than 9,000 miles.

353. Convergence

Consider the expression $\sqrt{2+\sqrt{2+\sqrt{2+\sqrt{2}}}}$ and prove that it converges to a limit of 4.

354. Circle and Point

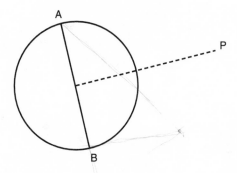

Prove that you can construct a perpendicular to any diameter of a circle from an outside point P, using a straight edge only.

355. Area of Triangle

Prove that the area of a triangle is the product of its base and half its height.

$$A = B \times \frac{H}{2}$$

9. The Mystery of the Fourth Dimension

Ours is a three-dimensional world, and such is the inadequacy of human imagination that most of us take for granted that three is the absolute maximum number of dimensions that can exist. We can visualize objects existing only in the first dimension—a line—or in the first two dimensions—a square, for instance—relatively easily, and we have no difficulty understanding the third dimension. Depictions of two- and three-dimensional figures have been the mainstay of every school of graphic art, from the earliest cave-drawings to Impressionism; while modern artists have shown themselves quite ready to work in 1-D, (no matter that its artistic possibilities seem rather limited to the more conservative art-lovers among us!).

Embedded in that very statement is an example of how at ease we are in the way we relate to the first three dimensions; when we look at graphic art, we instinctively fall in with the artistic convention that the pigment itself—the ink, paint, charcoal—has no dimension. If we did not, of course, every drawing would be three-dimensional; it would be impossible to depict a circle because ink *does,* in reality, have depth, and every circle ever drawn has actually been a cylinder.

However, the concept of dimensions beyond our own 3-D existence eludes us; we tend, therefore, to assume that they cannot exist. On page 123 I reproduce impressions of four-dimensional objects created by artists who have wrestled with the notion of dimensions beyond the third, though I doubt many readers will be able to visualize a fourth dimension even with the help of an illustration. There are those who have claimed they *can,* such as the nineteenth-century German physicist Hermann von Helmholtz (who maintained that the human brain was up to the task provided it was supplied with the necessary input data), but this transcendental feat seems beyond most of us.

Mathematics has no difficulty. It accepts the existence of dimensions beyond the third; it does not stop at the fourth dimension, in fact, but considers the possibility of dimensions existing through to the *n*th dimension as a matter of logical progression.

Imagine that your "mind's eye," as we call it, is a multidimensional slate on which a special drawing implement can leave impressions without excreting ink or any other drawing medium. First, make a *point;* a dot. This would be a *zero-dimensional* object. Then, from that point, move the drawing implement an inch to the right (or to the left). The impression it has now made, a line, is *one-dimensional.* Move the implement up or down an inch and you have a *two-dimensional* image. Completing the square by moving the implement an inch back along the horizontal and then an inch back along the vertical does not add to the number of dimensions. But by extending the square outwards or inwards an inch, you create a *three-dimensional* object, a cube.

Now imagine moving the cube an inch in the direction of a fourth dimension; what would be produced is known as a hypercube. "Tesseract" (after the Greek prefix, *tessera-,* meaning composed of four) is the generic term for any four-dimensional object.

Mathematicians and artists, and especially architects (in whom, it may be said, art and mathematics come together) have long been intrigued by the notion of a fourth dimension. The architect Claude Bragdon first published a drawing of a hypercube in 1913, and incorporated its design and other four-dimensional representations in his work. An example is his Rochester Chamber of Commerce Building.

As noted earlier, graphic artists and art-appreciators conspire in the fiction that there is no mass to the pigment. This being so, the media of graphic art—paper, canvas, whatever—are essentially two-dimensional media. You cannot draw *through* a sheet of paper, or paint *outwards* from a canvas. (Those artists whose style involves building up the pigment on canvas to create real depth do so for impressionistic effect—one could not, for instance, paint a life-size portrait and ever build up the pigment enough to make it a realistic 3-D portrait.) The way graphic artists realize 3-D images on 2-D media is by using *perspective* to imply their three-dimensional characteristics, and the key to making a four-dimensional, or an *n*-dimensional, drawing is simply the ability to draw perspectives of perspectives.

If perspective, which is employed in the illustrations of the hypercube and tesseract reproduced opposite, does not help most of us to visualize the fourth dimension, maybe an approach recently developed at Brown University will.

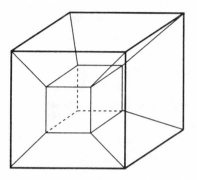

The tesseract

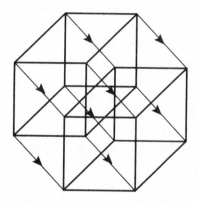

The hypercube by Claude Bragdon

With the aid of computer graphics, Thomas Bancroft (a mathematician) and Charles Strauss (a computer scientist) have produced continuous motion graphics of a hypercube moving in and out of a 3-D space, thereby cutting various images at different angles of the hypercube in a 3-D world.

Holograms have already become commonplace in the worlds of entertainment and advertising. The laser technology of the hologram offers a two-dimensional graphic medium that is capable of realizing three-dimensional images, but, unlike the archaic devices of paper and pen, holograms do not depend on the viewer going along with the illusion of perspective to depict the third dimension. The interplay of the different components of the laser produces a pattern which, when lit appropriately, *forces* the viewer to see 3-D, just as the few Hollywood movies shot in 3-D succeeded in transmitting thre dimensional images to audiences wearing the special spectacles provided. Perhaps, one day, 3-D holograms will be developed that can produce 4-D images that will finally force those of us not blessed with the special insight of a Bragdon, Strauss or Bancroft to "see" 4-D.

So far the thoughts presented in this preamble have been expressed before on many occasions and are little more than interesting speculative observations on this esoteric subject.

I now propose a novel approach consisting of a series of intellectual experiments leading to the compelling inference that a fourth geometrical dimension is likely to exist.

To this end let us assume that we are a team of scientists commissioned to investigate this subject and report our findings. We would have to begin by conceding that the sum of scientific knowledge to date has not produced any physical evidence of its existence, but perhaps we can, at least, adduce some intuitive indications that a fourth dimension *might* exist.

Imagine that somewhere in the universe is a 2-D planet—that is, a world whose inhabitants *believe* that their environment is only two-dimensional. How would we set about trying to convince a creature living there that his was actually a 3-D universe?

The equivalent of the Bancroft-Strauss sections (for example, a series of images of a sphere—a 3-D object—passing through a plane—the 2-D "world"—at different angles, as in the illustration below) might help 2-D man to wrestle with the concept of a third dimension, but it would certainly not prove its existence. After all, the sections would be no more than a collection of standard figures in 2-D geometry, namely circles.

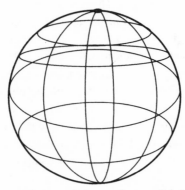

Different impressions made by a sphere as it intersects a plane.

Showing our 2-D man a drawing of a 3-D object would not help either. Just as, when we look at the perspective drawing of the tesseract, we cannot really "see" beyond the third dimension, 2-D man would not really be able to "see" 3-D. To him, a perspective drawing of a cube, for instance, might appear something like this:

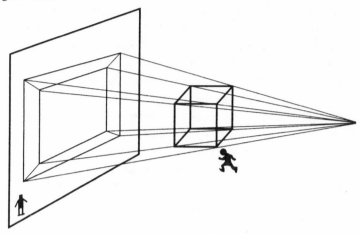

However, consider the following:

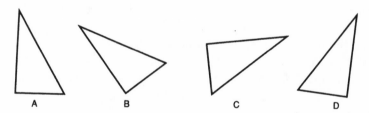

in the light of this definition of "Congruency" (from *The Penguin Dictionary of Mathematics*):

"Congruency describes two or more geometric figures that differ only in location in space. The figures are congruent if one can be brought into coincidence with the other by a rigid motion in space (without changing any distances in the figure). Note that two plane figures can be congruent without being identical. For instance, two scalene triangles with identical sides and angles are not identical if one is drawn as a mirror image of the other. They are, however, congruent (on this definition) since one can be rotated through 180° about an axis in the plane (or "picked up" off the plane and put down again the opposite way round). In the case of three-dimensional figures, this point is important since mirror images cannot be made coincident by a rigid motion in (three-dimensional) space. If two solid figures are identical, they are *directly congruent*. If each is identical to the mirror image of the other, they are *oppositely congruent*."

The previous illustration shows four right-angled triangles with all their sides and angles in common. They are, therefore, clearly congruent. But our 2-D friend, looking at the four triangles, is faced with a baffling phenomenon.

He can see that, by moving them around on the plane on which they presently lie, A, B, and C can be superimposed. They are *directly congruent* according to the terminology in the above definition. D, however, cannot, despite meeting the test of congruency with the other three, because it is *oppositely congruent*. No amount of shuffling around on the plane on which it lies will bring it in coincidence with the other three. The only way to achieve that is to pick D up, turn it around and then reintroduce it into the two-dimensional world. If our 2-D friend can only see that—if his imagination is up to it—then he can conceive of a 3-D world.

Now, some bright spark in our team of researchers might comment that, in our 3-D lives, we encounter similar phenomena: mirror images of irregular geometrical bodies that cannot be superimposed. Take, for instance, a human being's left and right hands (ignoring the minor differences which make them not quite perfect mirror images in practice). By definition, they are not directly congruent. Would a fourth dimension do the trick?

All very well, a doubter on the team might respond. Amusing, but hardly scientific. True, though this hint of the possible existence of a fourth dimension is no less intriguing for being anecdotal. But, in fact, our team can pursue its goal along a more scientifically orthodox path.

Modern cosmological theory views our universe as being in a state of continuous expansion. It is without beginning or end; as old galaxies die, new ones are formed out of matter created from nothing.

In 1922, the Russian physicist and mathematician Alexander Friedmann developed a concept of the universe based on certain assumptions which have been proven remarkably accurate by cosmological discoveries in the seventy years since. Friedmann's model postulated the following characteristics of the universe:

1. All galaxies are moving away from each other.

2. The farther apart they are, the faster galaxies move away from each other, the relative rate of acceleration being in proportion to the distances between galaxies.

3. There is no point in the universe that can be said to be the center of the expansion.

It is far beyond the scope of this book—and unnecessary—to detail the scientific evidence that led Friedmann to his conclusions. Suffice to say that observations based on the Doppler effect played an important part.

The Doppler effect is an observed change in frequency of light waves and sound waves as the source moves relative to the observer. The frequency of such waves increases—producing a move towards blue in the color spectrum and a higher pitch—as the object approaches, and decreases as it recedes. This effect is observable in everyday life; for instance, when a blowing car horn is passed on the highway.

The Doppler effect is used in astronomy to determine the relative velocity of different stars in our line of sight.

Many cosmologists have come to view our universe as a four-dimensional sphere (hypersphere) with a three-dimensional surface, having a circumference of the order of 100 billion light-years. One light-year is equivalent to 9,467,280,000,000 kilometers (5,917,050,000,000 miles). The suggested circumference of the universe is therefore approximately 9.467×10^{23} kilometers.

According to this model, what we perceive as straight, parallel lines are in fact great circles intersecting at two points at about 50 billion light-years distance, in each direction, on the hypersphere (in the same way that, on our globe, meridians that appear to the human eye as "endless" straight lines meet at the poles).

To help understand how this view of the universe has evolved, let us return to the world of our 2-D friend. Assume that his society has, over millennia, developed the same cosmological and other scientific theories we have. In early days, he would have believed his world to be a two-dimensional plane, the surface on which he walked being a straight line. Then 2-D Aristotle established that his world was shaped like a disc, though he still believed that 2-D World was stationary and the center of the universe.

Thanks to 2-D versions of Ptolemy (367–283 BC) and Copernicus (1473–1543), he learned more about the rotation of his world, and that it was not at the center of the universe, but merely one of several planets revolving around the 2-D sun; more recently, his understanding of some of the phenomena he encounters in his galaxy were refined by Kepler (1572–1630), Galileo (1564–1642) and Isaac Newton (1642–1727).

Not until the twentieth century did anyone imagine 2-D Universe to be continually expanding, or that the galaxy containing 2-D World could be just one of more than a hundred billion galaxies. Along came 2-D Friedmann with his model of the 2-D universe; suddenly, 2-D scientists found themselves unable to reconcile Friedmann's postulates with any structure known to 2-D physics.

Take, for instance, Friedmann's third postulate. In any plane expanding in all directions, there must be *one* point which will not move (and which, therefore, is the center of the expansion). Of course, there need not be any body at this center—it might simply be a notional locus—but some center there would have to be. (Friedmann did not claim merely that there was no galaxy at the center of the expansion; he determined that there was no center at all.)

In trying to reconcile 2-D Friedmann with established scientific wisdom, 2-D thinkers might be drawn to the same rational technique we are using here. They might imagine a society living in what it believed to be a one-dimensional universe (represented by a straight line):

1-D man would be baffled by *his* version of Friedmann, because it would seem to him that galaxies A and B, and any other galaxies in between, could only move apart if one point on the line AB remained static (i.e. formed the center of expansion).

But 2-D man would be able to tell him that Friedmann's third postulate works once one assumes a second dimension, thus:

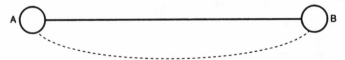

All points on the dotted line move outwards as it expands.

2-D sages would use this analogy to persuade their fellows that 2-D Friedmann's postulates betoken the existence of a third dimension. Suddenly, 2-D society would start to think of its universe not as a flat disc, but as a three-dimensional something. 2-D man would struggle to envision it, though we would recognize it readily as some form of expanding globe—a rubber balloon, say, slowly being blown up, with billions of dots painted on its surface to represent the galaxies.

Similarly, our team might conclude, the 3-D version of Friedmann and his postulates can only be made sense of if we consider our universe to be a four-dimensional something—however hard we may find it to visualize that something!

SOLUTIONS

i. Area of a Circle

Inscribe a polygon as shown and divide it into isosceles triangles.

Let G1, G2, G3 etc. be the bases (all equal) of the triangles, and H the height. The area of the polygon is then equal to:

$$\frac{H}{2}\,(G1 + G2 + G3 + \dots Gn)$$

Applying the Zero Option (see page xviii) by making the triangles' bases approach zero, the sum total of the bases becomes the circumference of the circle, and H becomes the radius R. Consequently, the above formula changes to:

$$\frac{R}{2} \times 2\pi R = \pi R^2, \text{ which is the area of a circle.}$$

ii. Steamship

Jim, standing on the port (west) side, is looking eastward, and John, standing on the starboard side, is looking to the west. In other words, the two sailors are facing each other.

iii. The Lion and the Unicorn

The only days on which the Lion can say "I lied yesterday" are Mondays and Thursdays. The only days on which the Unicorn can say "I lied yesterday" are Thursdays and Sundays. The only day on which *both* can say "I lied yesterday" is therefore a Thursday.

In case you need some help with the two earlier *liars and truthtellers* questions:

The one question which will identify which tribesperson is which is: "If I were to ask your friend what he is, what would he say?" A truthteller would answer "truthteller"; a liar would answer "liar."

The question which would have to be answered by a "Yes" is:

"Is your friend a liar?"

iv. The Two Friends

Let their present ages be X and Y respectively. Then we know that equation 1 is:

$$X + Y = 63$$

We also know that X is now twice as old as Y was (X – Y) years ago. Thus, equation 2 is:

$$X = 2[Y - (X - Y)]$$

Reduce to: $$X = 2(2Y - X)$$

and further to:

$$3X = 4Y$$

or: $$Y = \tfrac{3}{4}X$$

Solving the two equations yields: $X + \tfrac{3}{4}X = 63$

Therefore: $X = 36$ and $Y = 27$

v. Galileo's Paradox

The solution to this paradox is simple, and would have been obvious had the conclusory statement in the question not been (intentionally) a bit of a cheat.

The "honest" way of expressing the conclusion is to say that "a single point is equal *in area* to the circumference of a circle." Although the circumference of a circle may be thought to consist of an infinitude of points, this conclusion is not surprising when one remembers that both circumferences of circles and points have an area of zero.

vi. Hole in the Sphere

Without resorting to calculus, the problem can be solved as follows. Let R be the radius of the sphere. As the figure indicates, the radius of the cylindrical hole will then be the square root of $R^2 - 9$, and the altitude of the spherical caps at each end of the cylinder will be $R - 3$. To determine the residue after the cylinder and caps have been removed, we add the volume of the cylinder, $6\pi(R^2 - 9)$, to twice the volume of the spherical cap, and subtract the total from the volume of the sphere, $4\pi R^3/3$. The volume of the cap is obtained by the following formula, in which A stands for its altitude and r for its radius:

$$\pi A(3r^2 \times A^2)/6$$

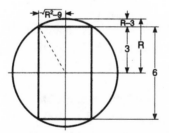

When this computation is made, all terms obligingly cancel out except 36π— the volume of the residue in cubic inches. In other words, the residue is constant regardless of the hole's diameter or the size of the sphere!

Using the Zero Option, make the diameter of the cylinder equal to zero. Then the volume remaining in the sphere is the whole sphere; i.e.:

$$\frac{4R^3}{3} \quad \text{Therefore,} \quad \frac{4.3^3\pi}{3} = 36\pi.$$

vii. Rope around the Equator

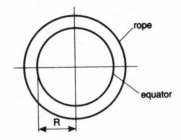

Let R be the radius of the equator.

Then the circumference   = $2\pi R$
The length of rope   = $2(R + 1)\pi$
                           = $2\pi R + 2\pi$

Deduct the circumference of the equator:

$$= 2\pi R + 2\pi - 2\pi R$$
$$= 2\pi$$

The Zero Option would have arrived at the same result at once. Assume the radius of the Earth, and therefore the circumference, to be zero. We are then left with a circle of 1 foot radius or 2π circumference.

viii. Gauss's Problem

Rather than add the numbers $1 + 2 + 3 \ldots 100$ consecutively, Gauss realized that the total was made up of 50 sums of 101, made up of the first and last numbers in the series, gradually approaching the middle of the series: $1 + 100$; $2 + 99$; $3 + 98 \ldots 50 + 51$. 50×101 yields the total, 5,050.

ix. Let A = B + C

When you factor $A(A - B - C) = B(A - B - C)$ you are in fact multiplying each side by zero, which automatically equalizes everything (by making it zero). While, on the face of it, the equation is correct, it is a fallacy to use it to prove that unequal numbers are equal.

x. Casino Game

This is a fallacy. Ignoring zero, the odds of either color coming up on any particular spin—be it the first or the sixteenth—are always 50/50. If 16 reds, say, have never come up in a row in the history of the casino, it is because 15 reds have only happened very rarely, and the history of the casino hasn't been long enough yet for 16 to have occurred. "Yet" is the operative word.

Just the play on which you've staked your shirt could easily be the one when the casino finally achieves a run of 16!

xi. Jack and John

Surprisingly, there is no fallacy in Jack's and John's reasoning. It is impossible for any balanced wager to favor both parties, but this is a very oddly constructed and *un*balanced wager. In the first place, in real-life wagers, the stakes are commensurate with risk—equal where the risk is equal; proportionate where the risk is unequal—but in this wager the risk is equal but the stakes are, *a priori,* not. Secondly, it is precisely the (unknown) relative values of their stakes that Jack and John are betting on. Thirdly, this wager is constructed so that the loser wins and the winner loses.

In these very unrealistic circumstances, it is indeed possible for both parties to feel that the wager is favorable.

xii. The Fax

It is not possible for any of the columns above the line to result in a carry-over of more than 1, which means that the M of MONEY cannot be a 2 or higher; in other words, M must be 1. For the SM column to produce a carry-over of 1, S must be a very high digit—8 or 9. In either case, the letter O must represent zero; SM can only total 10 or 11, but O cannot represent 1 since we have already determined that M does.

If O is zero, then the column E O cannot total as much as 10, meaning it cannot produce a carry-over. Therefore, S must represent 9 (not 8).

If O is zero, and no two letters represent the same digit, then N can only represent one digit higher than E (see the column E O N) and the sum of N R must produce a carry-over of 1. Thus:

$$(E + 1) + R (+ ?) = E + 10$$

The (+ ?) denotes the carry-over from the D E Y column; it may be zero (no carry-over) or 1. Looking at the above equation, we can confirm that the D E Y column *must* produce a carry-over, because, if it didn't, R would have to represent 9, and 9 is already represented by S. Now we know that R represents 8.

The D E Y column must total at least 12, since Y cannot be 1 or zero. Since neither D nor E can represent either 8 or 9, they can only represent 6 and 7 or 5 and 7 (any other two digits would not total 12 or more). Since $N = E + 1$, E must represent 5, N 6 and D 7. Then Y turns out to be 2.

xiii. Where There's a Will

This is an example of a trick *answer,* albeit quite a canny one.

The wise man temporarily added his camel to the 17, which apparently made it easy for the sons to divide up the camels according to the will without having to cut up a camel. The oldest son could take 9 of the 18 camels, the second 6 and the youngest 2, leaving one camel to be returned to the wise man.

The trouble is, of course, that this solution does not really divide the potentate's 17 camels in the specified proportions.

xiv. Number Series

According to the first letter of the number spelled, in reverse alphabetical order. Thus:

*Z*ero, *Tw*o, *Th*ree, *Six*, *Se*ven, *O*ne, *N*ine, *Fo*ur, *Fi*ve, *E*ight.

xv. Next

The next four numbers are: 1, 4, 1, 5, being the number of times a clock that chimes on the hour and the half hour chimes.

xvi. Move One Match

$$\sqrt{1} = 1$$

A bit of a cheat really, since the square root sign created by the solution is not quite accurate.

xvii. Artistic Fields

The usual method of solving such problems is to set up a matrix; indeed, matrices are often supplied in puzzle books offering identification problems:

	Dancer	Painter	Singer	Writer
Boronoff				
Pavlow				
Revitzky				
Sukarek				

Pavlow cannot be the writer or the painter, so we can place an X opposite his name in those two columns. Neither Boronoff nor Revitsky is the singer. Boronoff is also not the writer; neither is Sukarek.

Our matrix now looks like this:

	Dancer	Painter	Singer	Writer
Boronoff			x	x
Pavlow		x		x
Revitzky			x	
Sukarek				x

Clearly, Revitsky must be the writer. This means we can fill the other two boxes against his name with an X. Boronoff can't be the painter, because Revitsky has sat for the painter yet Boronoff has never heard of him; Boronoff must be the dancer. This makes Pavlow the singer and Sukarek the painter.

xviii. The Joker

When you remove one card from the group of three, the probability that the joker will be your card is 1/3 and the probability that it will be one of the two left on the table is 2/3. However, since it is certain that one of the cards left on the table is an ace, the odds are not altered by your friend turning the ace over. Which means that the odds of the card left face down on the table being the joker are still 2/3.

1. Tangent Triangle

The answer is 20 units. Since lines tangent to a circle from an exterior point are equal, YA = AP and BP = BX. Since the side AB of the triangle equals AP + BP, the perimeter of the triangle must be 10 + 10 = 20 units.

This is one of those curious problems that can be solved by the Zero Option. Since P can be anywhere on the circle from X to Y, we move P to a limit (either Y or X). In both cases one side of △ ABC shrinks to zero as side AB expands to 10, producing a degenerate straight-line "triangle" with sides of 10, 10, and 0 and a perimeter of 20.

2. Area of Overlap

First, extend two sides of the large square as shown in the diagram (below) by dotted lines. The small square has now been divided into four equal parts. Since the total area of the small square is 9 inches, the area of the overlap must be $\frac{9}{4}$, or 2.25 square inches. Note that the information given about the lengths of AB and BC was not required for solving this problem. In fact, the area of overlap will be constant regardless of the position of the large square as it rotates around D.

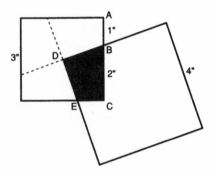

3. The Pond

The largest (and only) ellipse is the border of the pond which encloses an area of 50 square feet. The other two ovals are not ellipses.

4. Two Angles

I made $\angle$ B first, and constructed $\angle$ A three times as large.

5. Area of Annulus

The center of the circle and points A and B form a right-angled triangle. Let r_1 = radius of inner circle and r_2 = radius of outer circle. From the Pythagorean theorem, $r_2^2 - r_1^2 = 1$. The area of the annular region is $\pi(r_2^2 - r_1^2)$. Since the term in parentheses equals unity, the area of the annular region is π^2 square inches.

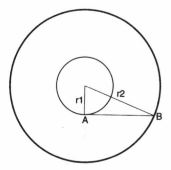

Now try the Zero Option: Let the radius of the inner circle be zero. Then the area of the annulus becomes a circle with area π^2 inches.

6. The Round Window

Since there are 8 outer panes, each of which is to have the same area as the inner pane, the area of the whole window must be 9 times the area of the inner pane. Now the area of a circle is proportional to the square of its diameter; hence, if the area of the whole window is 9 times that of the inner pane, the diameter of the whole window is 3 times the diameter of the inner pane, or 6 feet. Hence, the spokes should be 2 feet or 24 inches long.

7. The Hypotenuse

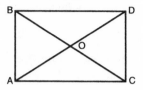

If ABC is a right-angled triangle with BC the hypotenuse, the rectangle formed by ABCD in the figure above will have two diagonals, AD and BC, meeting at O. O is the mid point of both BC and AB, and hence the length AO must be equal to half BC.

8. Hex Sign

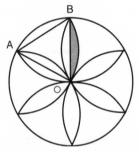

Let A and B be the end points of two neighboring petals, and O the center of the hex sign. Then AB = OA = OB = 1 inch.

We observe that half of one of the petals is equal to the difference between △ ABO and the segment of a circle (see figure above).

The triangle is an equilateral triangle with sides equal to 1 inch. The area of such a triangle can be calculated by the formula $\frac{1}{2}$ Base × Height.

Dropping a perpendicular line from one of the apexes to the middle of one of the sides, the height can be readily ascertained by the theory of Pythagoras to be:

$$\sqrt{(1)^2 - (\tfrac{1}{2})^2}$$

and since the base is 1 inch, the area is

$$\tfrac{1}{2}\sqrt{1-\tfrac{1}{4}} \; - \; \tfrac{1}{4}\sqrt{3} \; = \; 0.433 \text{ square inches}$$

Turning to the segment of the circle in the right-hand illustration below, the angle at A is 60°, which is one sixth of 360°.

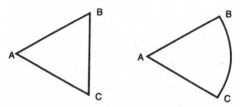

Therefore, the area of the segment is one sixth of the area of a circle with 1 inch radius or

$$\frac{\pi}{6} = 0.524 \text{ square inches}$$

The difference between these two areas is the area of half a petal.

Therefore, half a petal = 0.524 − 0.433

$$= 0.091 \text{ square inches}$$

and the area of a whole petal is 0.182 square inches.

9. How Large is the Cube?

The surface area of the cube is 6 times the area of one of its 6 faces. Suppose the cube has an edge x inches. One of its faces has an area of x^2 square inches. So its total surface area is $6x^2$ square inches. But this must be equal in number to its volume or $x \cdot x \cdot x = x^3$ cubic inches. So $6x^2 = x^3$, which means $x = 6$. Therefore the cube has a side of 6 inches.

10. As the Fish Swims

Did it seem as if there wasn't enough information given to solve this problem? There was—if you remembered the rule that when a right triangle is inscribed in a circle, its hypotenuse is the circle's diameter. The pool's diameter is 100 feet—the hypotenuse of a right triangle having sides of 60 feet and 80 feet. (The square of the hypotenuse equals the sum of the squares of the other sides.)

11. Pentagram

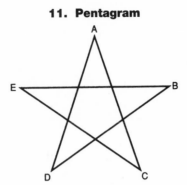

Let x be $\angle$ DAC in the figure above. The line AD must be turned through $\angle$ x to bring it to AC. AC must be turned through x again to bring it to EC. Again EC must be turned through x to bring it to EB. Proceeding in this manner, we find that AD must be turned through $\angle$ x 5 times to bring it back to itself, but in the process AD has acquired an opposite sense, so that A is now where D was and D is now where A was. Hence, the line AD has been turned through 180°. From this we calculate $5x = 180°$ or $x = 36°$.

The angles at the points of the star are 36°.

12. Diagonal Problem

Line AC is one diagonal of rectangle ABCD. The other diagonal, BD, is the radius of the circle, which is 10 units. Since AC = BD, line AC is also 10 units long.

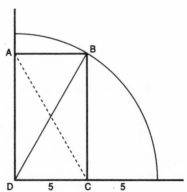

The Zero Option also works. AC is obviously the same length irrespective of C's position. Let C coincide with D, thus aligning AC with the radius.

13. Squares from a Square

By connecting the midpoint of each side with the corner opposite, we get 5 small squares, as shown.

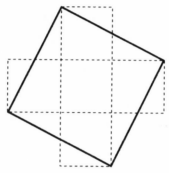

You will see from the diagram that the areas of the small squares add up to that of the large one. The triangles outside the square are congruent to the triangles inside the square.

14. The Goat

$\frac{3}{4}$ of original circle	=	1039.5	square feet
$\frac{1}{4}$ of circle, radius 14 feet	=	154.0	square feet
$\frac{1}{4}$ of circle, radius 7 feet	=	38.5	square feet
		1232.0	square feet
Original circle	=	1386.0	square feet
Grazing $^{1232}/_{1386}$	=	88.8% × $100	
Should pay		$88.80	
Reduction		$11.20	

15. Water Lily

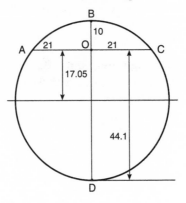

17.05 inches

Using theory of chords:

i.e. $\text{AO} \times \text{OC}$ = $\text{BO} \times \text{OD}$

∴ 21×21 = $10(2x + 10)$

Or 441 = $20x + 100$

∴ x = 17.05

Or, if you are unfamiliar with the theory of chords, you use the theory of Pythagoras:

$$x^2 + 21^2 \ = \ (x + 10)^2$$

$$x^2 + 441 \ = \ x^2 + 20x + 100$$

or $20x$ = 341

∴ x = 17.05

16. The Sphere

17.146 centimeters.

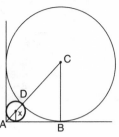

$$AC = \sqrt{2}$$

$$AD = \sqrt{2} - 1$$

$$X = AD - X.\sqrt{2}$$

$$X + X.\sqrt{2} = \sqrt{2} - 1$$

$$X\left(\sqrt{2} + 1\right) = \sqrt{2} - 1$$

$$X = \frac{\sqrt{2} - 1}{\sqrt{2} + 1}$$

17. Division of Land

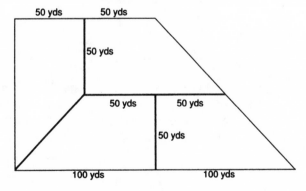

18. Center of a Circle

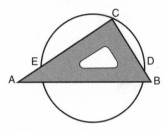

Place C, the right angle of the draughtsman's triangle, on the circumference, as shown above. D and E, where the triangle's legs cross the circumference, are the endpoints of a diameter. Draw it, and get a second diameter the same way. Their intersection is the center of the circle.

19. The Window

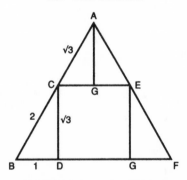

By similar triangles, since CEGD is a square, CE is parallel to BF, and CAE is equilateral, AC = CE = DC. Draw AG as a perpendicular bisector of CE; by symmetry GC = ½CE = ½AC, by similar triangles CB = 2BD. Let BD = 1, then BC = 2 and CD = 3 = AC, by the Pythagorean theorem. Then if AB = 1,

$$CD = \frac{\sqrt{3}}{2+3}$$

20. The Arch

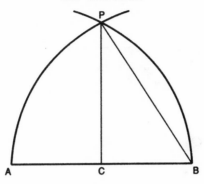

Bisect AB at C, and join PB and PC. PB is the radius of arc AP, as is AB. By the theory of Pythagoras:

$$PC^2 = PB^2 - CB^2$$
$$= 1^2 - (\tfrac{1}{2})^2$$
$$= 1 - \tfrac{1}{4}$$

Therefore, $PC(h) = \sqrt{\dfrac{3}{4}} = \dfrac{\sqrt{3}}{2}$

21. Equal Areas

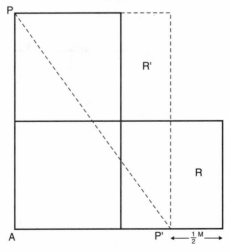

The rectangle R has been transposed to R′ and the dividing line PP′ is the required one. The transposition has no effect on the areas involved. The proof is evident.

There is also a simple mathematical solution: the rectilinear right-angled figure has an area of 3 square inches. We are therefore trying to find point X so that triangle APX = 1.5 square inches.

Therefore $\dfrac{PA \times AP'}{2} = 1.5$

or $\dfrac{2 \times AP'}{2} = 1.5$

Therefore $AP' = 1.5$

22. Angular Problem

Here is one of numerous possible solutions. I think it's the best.

Construct the additional squares shown in the diagram below. ∠B is equal to ∠D because they are corresponding angles of similar right-angled triangles. ∠A plus ∠D are equal to ∠C. ∠B is equal to ∠D. Therefore ∠C is the sum of ∠s A and B.

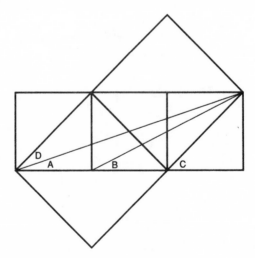

23. Overlapping Circles

Three equal circles, each passing through the centers of the other two, can be repeated to form the pattern shown in the figure below.

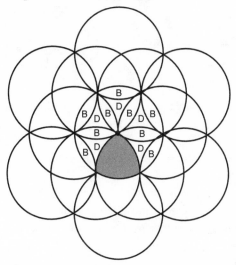

Each circle is made up of 6 D shapes and 12 B shapes. One quarter of the area of a circle must therefore be equal to $1\frac{1}{2}$ Ds plus 3 Bs. The area common to 3 mutually intersecting circles (shown shaded in the diagram) consists of 3 Bs and one D. Therefore it is smaller than a quarter of the area of a circle by an amount equal to half a D. In fact, the mutual overlap is just over 0.22 of the circle's area.

24. Sideways

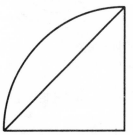

25. Three-dimensional Object

The front and side views are shown along with the object below:

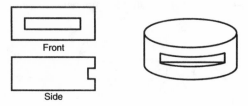

There are other more complex shapes that could produce the two views shown. The most common suggestion is a rectangular box with a "drawer" hole through it and notches at the back end. (Such an object would have been rotated 90° to the left rather than to the right, to get the side view pictured.) Or the shape could be rounded on the notched side only. In an informal survey, fewer than 20 percent of the people who came up with a workable answer thought of the simplest solution: the object is a square cylinder with a notch in it. From directly above, the cylinder appears as a perfect circle.

26. Circles in the Squares

Here's another problem that doesn't seem to provide enough information. The key here is to avoid getting bogged down in calculating the size of each individual square and circle, and to attack the problem as a whole.

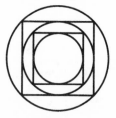

If you set up the problem in the way most people do, you may first calculate the diameter of the medium-sized circle: it is equal to the side of the larger square.

Since the side of a square is equal to its hypotenuse divided by the square root of 2, and since the hypotenuse of the large square is 10 inches, the diameter of the medium-sized circle must be 10 divided by the square root of 2. If you don't know what the square root of 2 is, don't worry, because the same calculation must be performed again to find the diameter of the small circle.

Since 10 inches must be divided by the square root of 2 twice, it is the same as dividing by 2 only once. So the diameter of the inside circle is 5 inches. You've got away with solving a problem using the square root of 2, without having to know what it is!

I said that this is the way most people set up the problem; and, as you can see, it can be solved this way, in your head. But if you can break away from the typical way of doing things there's an even simpler solution. Most people orient the two squares the same way, but nothing in the rules of the problem demands this. If the inner square is rotated 45° and stood on its corner, as shown below, there is a way you can calculate the diameter of the inner circle without even considering the square root of 2. What is the quick insight that solves this version of the problem?

When the problem is set up as shown below, you see immediately that the radius of the large circle (AB), which we know to be 5 inches, is equal in length to a side of the inner square (CD) because both lines, AB and CD, are diagonals of the same square. Since the small circle is inscribed in the inner square, it has a diameter that equals the side of that square (CD) or 5 inches.

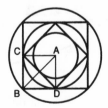

Another way of putting this is that, since the two diagonals of square ADBC bisect each other, the radius of the small circle is half the length of line AB, which is the radius of the large circle.

27. Spider and Fly

The shortest route the spider can walk to get to the fly is 40 feet. This is shown in the diagram of the unfolded room below. It is interesting to note that this route involves the spider walking across 5 of the room's 6 sides.

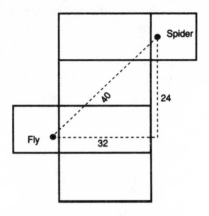

28. Overlapping Areas

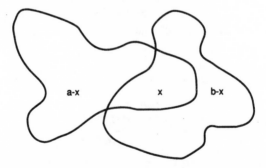

In general, if any two areas a and b have a common area x, then the non-overlapping portions are $a - x$ and $b - x$. The difference of the areas of the non-overlapping portions is $a - b$. In this problem, the difference is $\pi(20)^2 - \pi(15)^2$ or 175π.

29. The Vanishing Square

The forger took 14 $20 bills and cut each of the bills into two parts as shown in the figure below (along the dotted lines).

He then stuck the upper section (using adhesive tape) to the appropriate section of the next note, with the result that 14 bills became 15. Each note was of course shortened by one-fifteenth of its length, which was not noticeable.

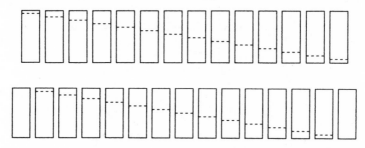

30. The Hula Hoop

Since motion is relative, consider the hoop as fixed and the poor girl whirling around. The original point of contact on the girl traverses the diameter of the hoop twice, and this is the required distance.

As the girl whirls, the original point of contact C on her waist traverses the diameter BD, since

$$\text{arc } AB = R\theta = (R/2)(2\theta) = \text{arc } AC$$

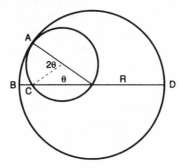

The answer can be made clearer with the illustration as follows:

(1) Arc AB = arc AC if there is no slippage. Clue that the girl's waist is not smooth.

(2) Side motion is relative, consider the hoop as fixed and the girl whirling about.

(3) Point C travels along BD twice; i.e. a distance of twice the large diameter or 4 times the small diameter = perimeter of the square circumscribing the girl's waist.

31. Bisect a Line

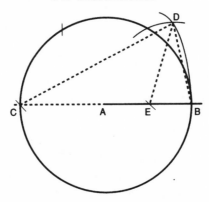

Describe a circle with radius AB and A as its center, and an arc with the same radius and B as its center. With the compasses still set at radius AB, start from B and make 3 successive arcs on the previous arc (center A) to locate point C. With radius CB and C as the center, describe an arc cutting the previous arc (center B) at point D. Finally, with the original radius AB and D as the center, make an arc which will intersect AB at its midpoint E. The broken lines in the diagram are used in the proof only, which is: "Point C is on the extension of line BA because 3 successive radii subtend a semicircle, and △s BCD and EDB are similar because they are isosceles with a common base angle. Since CB is twice BD, EB must then be half BD, or half AB."

32. Centers of Three Circles

1. Yes. Try it and see.
2. No, because any three points (centers of the circles) always make a triangle, but any three straight lines only if none of the lines is greater than the sum of the other two.

33. The Magnifying Glass

I drew an angle.

34. Isosceles Triangle

With one 10-inch side as base and the other 10-inch side free to rotate (as shown above), the triangle's area is greatest when the altitude is maximum. The third side will then be $10 \times \sqrt{2}$.

35. Two Hexagons

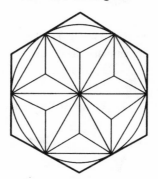

Instead of inscribing the hexagon as shown in the question, turn it as shown above. The lines divide the larger hexagon into 24 congruent triangles, 18 of which form the smaller hexagon. The ratio of areas is therefore $18:24 = 3:4$.

36. The Flag

$h = 2$ feet

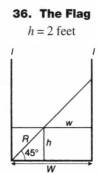

In the figure above, the vertical lines, ll, represent the available space with width = 4 feet. The red square, R, is put at one side. Since R is always square, the 45° diagonal is the locus of its free corner, and the rectangle to its right, with height h, and width w, is the total area of white left.

Change scale, making $W = 2$: when $n = w = 1$, then the area of white equals 1. If h and w are unequal by the amount $2n$, area $= (1 + n) \times (1 - n) = 1 - n^2$, which is less than 1; therefore the maximum area is obtained when $h = 2$ feet in the original scale.

There is another intuitive proof: w and h (always equal to w) form rectangles with area w and h. It is axiomatic that the area is maximized if $w = h$ (a square).

37. Tangent Circles

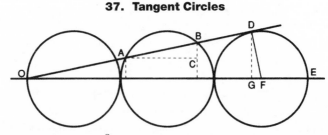

The length AB is equal to $\frac{8r}{5}$. Determine the equation of the tangent OD and the equation of the middle circle. Solve these equations simultaneously to find the points of intersection A, B. Then compute the distance between these points.

Take O as the center of rectangular coordinates, with OE as the x-axis. The equation of line OD is then given by the ratio of DG (perpendicular from D to OE) to OG. Since ODF is a right triangle, the altitude DG is equal to $\frac{(OD)(DF)}{OF}$.

DF equals r, and OF equals $5r$. From these values, the other terms can be computed. The equation of OD is found to be:

$$\frac{y}{x} = \frac{\sqrt{6}}{12}$$

The equation for the middle circle is:

$$(x - 3r)^2 + y^2 = r^2$$

Solve the two equations simultaneously for the value of x, which is found to be:

$$x = \frac{72r}{25} + \frac{8r}{25}\sqrt{6}$$

The two values of x are the abscissae of the points B and A.

The difference between these values, equal to

$$\frac{16r}{25}\sqrt{6}$$

is the length of AC. Since ABC is a right triangle by construction, AB can be computed from AC and BC. The latter can be computed by use of the equation for OD. The desired length AB is found to be $\frac{8r}{5}$.

38. The Avenue

The area of the two triangles is $\frac{5}{12}$ of the area of the estate. Therefore the shortest side of each triangle is $\frac{5}{12}$ of the side of the square. Hence the longest side of each is $\frac{13}{12}$ of the side of the square, i.e. of 1,430 yards. The central avenue is, of course, the same length. So the central avenue is 1,430 yards long.

39. Four-wheel Cart

For the outer wheels to go twice as fast as the inner wheels, the outer circle must have twice the circumference of the inner circle. Therefore the 5 feet between inner and outer wheels must be half the radius of the outer circle, giving the outer circle a diameter of 20 feet and a circumference of 20π, or 62.832 feet.

40. Crossing the River

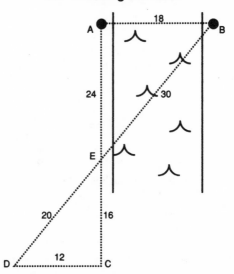

Measure any convenient distance along the bank from A to C, say 40 yards. Then measure any distance perpendicularly to D, say 12 yards. Now sight along DB and find the point E.

You can then measure the distance from A to E, which will here be 24 yards, and from E to C, which will be 16 yards.

Now AB : DC = AE : EC, from which it is evident that AB, the width of the river, must be 18 yards.

41. Wheels and Spheres

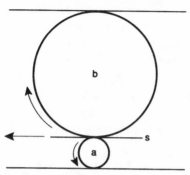

Neither sphere would go ahead of the other. Since the top surface of the lower sphere is in non-skidding contact with the bottom surface of the top sphere, it is possible to imagine a sheet of paper, s, between them, moving to the left. When s moves x inches, the center of the bottom sphere, a, moves $\frac{1}{4}x$ inches in the same direction. The center of the top sphere, b, moves likewise. So a remains vertically below b regardless of the direction of motion.

42. The Shed

Since the view of the roof is an equilateral triangle, the altitude of each triangular face, a, must be equal to the width of the shed, 10 feet, as shown in the left-hand figure. This altitude is twice that of each of the four triangles that would make up the plan of the 10 feet × 10 feet (or 100 square feet) plan of the shed, shown in the right-hand figure. When the altitude of a figure is doubled, the area is doubled, so the area of the roof is 200 square feet.

43. Two Gold Coins

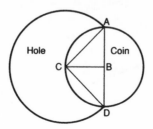

When the smaller coin reaches the point where it tips into the hole, its diameter AD is a chord of the hole. C is the center of the hole, and B is the center of the coin. CB and AD are at right angles. AC and DC are radii of the hole. AB and BC are radii of the coin. Let the length of AB = 1 unit. Since AB = CB, the length of CB also = 1 unit. The length of AC is thus $\sqrt{2}$. Since the coins are both the same thickness, their weights are in proportion to their areas, which in turn are in proportion to the square of their radii. Hence the bigger coin weighs twice as much as the small one (i.e. $1:(\sqrt{2})^2$ or $1 : 2$).

44. Pieces of Paper

The fifth set is the four points of intersection of the circle E, which needs no proof.

The sixth set is points A, B, C and D. ∠s DAB and BCD are right angles. If line DB is drawn, we see that points A and C are both subtended at right angles from DB. Imagine DB as the diameter of a circle. Any points subtended at right angles from the diameter of a circle will lie on the circumference of that circle.

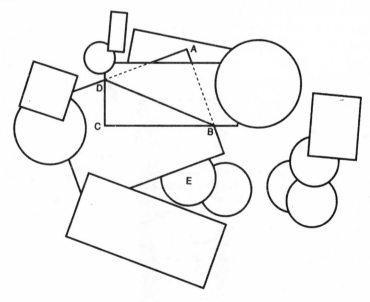

45. Balloon and Box

As the balloon sank and fell, it described a hyperbola. The center of the balloon, C, moved so that its distance from the edge, E, was always equal to the radius of the balloon. At any given position, its distance from the center of the hemicylinder was always equal to the radius of the balloon plus the radius of the cylindrical lid.

46. Find the Area

Each sector includes an angle of 60° or $\frac{1}{6}$ of a circle with radius of 1 foot. The three sectors have an area of $\frac{\pi}{2}$. Deduct this from the area of the triangle, leaving:

$$\sqrt{3} - \frac{\pi}{2}$$

47. Self-congruent

The third type of self-congruent line is called a circular helix—a line that spirals with a constant angle around a cylinder that has a circular cross-section. As can be seen from the figure below, any portion of the helix will fit any other portion.

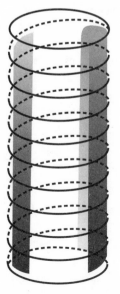

48. Two to One

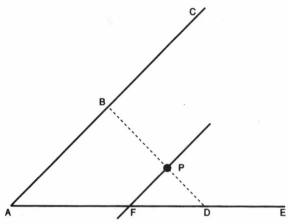

Find point F, so that AF = 2 × FD. Draw a parallel to AC through P. △s FPD and ABD are similar. As AD : FD = 3:1, the same ratio applies to BD and PD. Therefore BP = 2 × PD.

49. Trisecting the Square

The trisecting lines also trisect each side of the square.

This is easily seen by dividing any rectangle into halves by drawing the main diagonal from the corner where the trisecting lines originate. Each half of the rectangle obviously must be divided by a trisecting line into two triangles, such that the smaller is half the area of the larger. Since the two triangles share a common altitude, this is done by making the base of the smaller triangle half the base of the larger.

50. The Circular Table

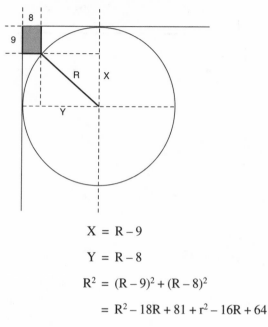

$$X = R - 9$$

$$Y = R - 8$$

$$R^2 = (R - 9)^2 + (R - 8)^2$$

$$= R^2 - 18R + 81 + r^2 - 16R + 64$$

or $$R^2 - 18R + 81 - 16R + 64 = 0$$

∴ $$R^2 - 34R + 145 = 0$$

Solve the quadratic equation:

$$R^2 - 34R + 289 = 144$$

$$(R - 17)^2 = 144 \qquad\qquad R = 29$$

There is a second solution obtained by inscribing the circle in the shaded rectangle, which yields a radius of 5 inches. This, however, cannot be described as a table.

51. Up the Garden Path

The area of the path is $66\frac{2}{3}$ square yards.

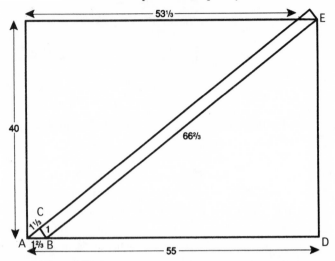

Note: The width of the path is not to scale for the sake of clarity.

$\triangle$ABC is similar to $\triangle$BED. Also AB + BD = 55. From this $\overline{BD}$ is calculated to be $53\frac{1}{3}$.

Also:

Let the area of the path be "X."

Then $40 \times 53\frac{1}{3} + x = 2200$ (area of garden).

or $x = 2200 - \dfrac{6400}{3}$

$\therefore$ $3x = 6600 - 6400$

and $x = 66\frac{2}{3}$

52. Four Towns

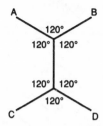

The shortest solution is 27.3 miles, as shown in the diagram above.

53. Divide the Circles

By superimposing C on B, you can mark the four equal areas as shown.

Proof: the four circles have a total area of 50π square centimeters. Consequently each of the equal units should be 12.5π square centimeters. Each half of circle A meets the condition, and the two other areas, as marked, are equal and must therefore have an area of 12.5π square centimeters each.

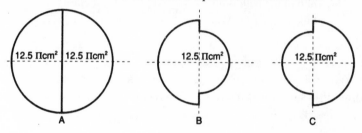

54. The Spiral

Draw a line and select any two points, A and B, 5 millimeters apart. Describe semicircles on the line, alternately using A and B as centers, and joining the ends accurately.

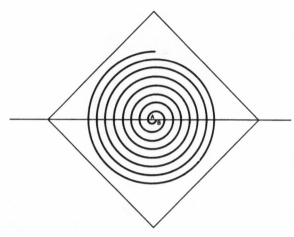

This is not a true geometrical spiral, but it is a spiral, "as shown."

55. The Banner of St. George

The answer given in *Amusements in Mathematics* by H. E. Dudeney (Dover Publications) is as follows:

As the flag measures 4 feet by 3 feet, the length of the diagonal (from corner to corner) is 5 feet. All you need do is deduct half the length of this diagonal ($2\frac{1}{2}$ feet) from a quarter of the distance all round the edge of the flag ($3\frac{1}{2}$ feet)— a quarter of 14 feet. The difference (1 foot) is the required width of the arm of the red cross. The area of the cross will then be the same as that of the white ground.

Proof: Let the width of the cross be x.

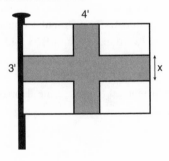

Then:

$$4x + 3x - x^2 = 6 \text{ or}$$
$$x^2 - 7x + 6 = 0$$

Solving the quadratic equation:

$$x^2 - 7x + 12.25 = 6.25$$
$$(x - 3.5)^2 = 6.25$$
$$x - 3.5 = \pm 2.5$$

As x cannot be 6, it must be 1 foot.

56. Inner Space

Down a mineshaft, the gravitational effect would be less because some of the Earth's mass is above you. That mass pulls you up, so it cancels the effect of some of the mass below your feet that pulls you down.

The point of this question is to show that gravity does not come from one source of attraction at the center of the Earth, as many people suppose, but comes from the combined attraction of all the parts of the globe working together. Our idealized Earth of "uniform density" is fictional, of course, and in practice gravity would increase slightly as one got closer to the Earth's dense nickel-iron core. But as one descends into the core, the effect of gravity would be less and less, and at the Earth's center it would be zero.

57. Weigh-in

The fluctuations result from the up-and-down movement of your blood's center of gravity as your heart goes through its pumping cycle. For a person weighing 165 lb, there is a fluctuation of about 1 ounce with every heartbeat.

58. The Rifle

Both bullets will hit the ground at the same time. Horizontal velocity does not affect downward acceleration.

59. Cracked Up

The thick drinking glass will break first. Glass is a poor conductor of heat. In the thin wine glass, heat passes quickly from the inner to the outer surface of the glass, and the glass expands with relative uniformity. When you pour hot water into the thick glass, its inner surface expands quickly while its outer surface remains the same size; this results in enormous stress on the glass, and so it cracks. That's why you can safely pour hot Irish coffee into a delicate wine glass—but don't try the same thing with a tumbler.

60. Roll-off

The ball will reach bottom first because it has less of its mass on the outside, which gives it less angular (turning) inertia. The ring rolls most slowly because, with its mass distributed away from the center of rotation, it has more inertial resistance to turning. For similar reasons, a figure skater can spin faster when her arms are tucked in close to her body.

61. Mountain Time

A mechanical watch runs faster in the mountains. The principal reason is the low atmospheric pressure and air density at high altitudes; the balance wheel has less air to push around, so it oscillates a bit faster.

Some infinitesimal effects may be predicted from the theory of relativity and will operate on electronic watches as well as wind-up ones. Einstein's theory says that time moves more slowly as (a) gravity increases and (b) velocity increases. The lessened gravity in the mountains would make a watch run relatively faster. This effect would be counteracted very slightly by the fact that, in the mountains, you're farther from the Earth's axis of rotation and therefore traveling through space at a slightly faster rate, which would slow time.

62. Launch Pads

Spacecraft are launched in the direction of the Earth's rotation, from west to east. To take advantage of the maximum "push" from the Earth, launch sites are located as close to the equator as possible.

63. Poles Apart

Antarctica is a continent, and land is a poor heat conserver, so Antarctica's ice doesn't really melt readily. The Arctic is over an ocean, which has a high heat capacity. It warms the surface, melting much of the Arctic's ice.

64. Hard Skate

It's harder to ice skate when the air temperature is very cold. We tend to think of ice as inherently slippery, but it isn't. When you skate, the ice beneath your skates' sharp runners melts temporarily, creating a thin lubricating film of water. When it is very cold, the ice does not melt readily; this makes the ice feel "sticky" and harder to skate on.

65. The Suspended Egg

First fill the glass half full of water and dissolve enough salt in it to make it dense enough for the egg to float on top. Then add more unsalted water to the top, filling the brim.

66. Two Bars of Iron

Take either bar and push one end against the middle of the other bar, forming a T. If the magnetized bar is the top of the T, there is no pull on the other bar.

67. The Bird Cage

If the bird is in a completely airtight box, the weight of the box and the bird will be the same whether the bird is flying or perching. If the bird is flying, its weight is borne by the air pressure on its wings; but this pressure is then transmitted by the air to the floor of the box. If the bird is flying in an open cage, part of the increase in pressure on the air is transmitted to the floor of the cage, but part is transmitted to the atmosphere outside the cage. Hence the cage with the bird will be lighter if the bird is flying.

68. Exception to the Rule

Water expands below 3.98°C (39.164°F) down to 0°C (32°F).

Otherwise, ice would be heavier than water, with catastrophic consequences. For instance, lakes and oceans would freeze from the bottom up, destroying marine life as we know it.

69. Water Level Problem

The water level neither rises nor falls: it stays the same. The reason an ice cube floats is because its volume has expanded during crystallization. Its weight remains the same as the weight of the water that formed it. Since a floating body displaces its weight, when the ice cube has melted it will provide the same amount of water as the volume of water it displaced when it was frozen.

70. Boat in the Bath

The nuts, bolts and washers in Rupert's boat displace an amount of water equal to their weight. When they sink to the bottom of the bath, they displace an amount of water equal to their volume. Since each item weighs considerably more than the same volume of water, the water level in the bath goes down after the cargo is dumped.

71. Space Station

Zero gravity would prevail at all points inside a hollowed-out asteroid, so it would float permanently at the same location.

72. Bird on the Moon

The bird couldn't fly at all on the moon because there is no lunar air to support it.

73. The Goldfish

The scale registers an increase in weight equivalent to the amount of liquid displaced by the suspended goldfish.

74. Speed of Sound

The speed of sound remains at 740 miles per hour; it does not get an extra "push" by approaching you. The sound waves will be crowded closer together, however, resulting in a higher pitch, known as the Doppler effect.

75. Two Bridges

The smaller bridge is twice as strong. If a steel girder, B, is twice the size of girder A in every dimension, it will be twice as strong as girder A but it will weigh eight times as much. The double-sized bridge could be so weak that it would collapse under its own weight.

76. Two Sailboats

Neglecting their weight, big sails are just as strong as small ones. The reasoning of the previous problem doesn't apply in this case, because as the size of a sail grows the force of the wind against the sail grows at the same rate.

This is an example of why physicists are so fond of using words like "neglecting" and "roughly." In practice, the weight of the sail has to be taken into account; similarly, since winds 20 feet up are liable to be fiercer than winds only 10 feet up, a larger sail will be made of stronger fabric than a small sail. The point of the question, in contrast to the previous one, is that the load on a sail (in the form of wind pressure) does not increase as the cube of the linear dimension, as the load does on a bridge (in the form of weight).

77. Aircraft Temperature

The pressurized cabin in an airliner keeps air compressed to sea-level pressures. At a high altitude, this would raise the cabin temperature to 130°F or higher if air-conditioners were not used to extract heat from the air.

78. Fan Power

Strange as it seems, the fan will propel the boat—backwards! The reason is that not all of the wind generated by the fan is caught by the sail. The forward action is not enough to counter the backward reaction, so the boat is propelled backwards.

79. Flags

The flags show a star shining between the horns of a crescent moon. Since this area is merely the unlit portion of the moon, any stars that might be in that part of the sky would be hidden from view.

80. Where Are You?

Try to spin one of your coins on the floor of your room: the coin will refuse to spin. By conservation of angular momentum, a spinning object tries to maintain its position in space. Since the spinning station is continually changing your position in space, a coin that is spun will keep changing its orientation, to correct its angular momentum, and will topple and fall.

81. Bridge Towers

To compensate for the curvature of the Earth.

82. Drops and Bubbles

The bubbles would move towards each other. If water is removed from one spot in all space (bubble A), the gravitational balance surrounding it is upset, and the net effect on a nearby molecule of water is that it is drawn towards greater mass; that is outwards, away from the bubble. If there are two bubbles, the water between them acts as if it is repelled from both, and the bubbles would move towards each other.

83. Special Sphere

Since a 20-foot-diameter sphere of 300 lb would weigh less than an equal volume of air (about 330 lb at sea level and 32°F, given pressure of 0.081 lb per cubic foot with air density at 14.7 lb per square inch), the sphere would rise into the air. It would level off at several thousand feet and could remain there for years, riding the winds.

84. Saving Energy

The method will work. You can boil water at a low temperature, but would you want to? If you brought a pot of water to a rolling boil at, say 160°F, food would take for ever to cook in it, and water at the temperature wouldn't kill bacteria either. It is because of the effect of atmospheric pressure that some recipes and frozen food instructions recommend boiling food for a longer time if it is being cooked at a high altitude. Indeed, increasing the pressure is desirable, which led to the invention of the pressure cooker.

85. Shut That Door

Nothing would happen because there is no oxygen in the room.

86. The Telephone Call

Father and child; or paternal uncle/aunt and nephew/niece.

87. Relations

Jill is Jean's mother; alternatively, Jean is the daughter of Jack's wife's brother or sister.

88. Sisters

They are two of a set of triplets or quadruplets, etc.

89. How Close?

Father.

90. The Painting

The man's own son.

91. Sons' Ages

There are only fourteen combinations of ages that satisfy the first clue, even allowing for two of the sons to be twins. (They can't all be triplets if their ages, in whole years, add up to 13.) The second man knows his own age, and yet the second clue doesn't give him the answer, so his age must be 36—the only product that occurs twice among the fourteen possible combinations.

The third clue indicates that there is only one oldest son, not two, ruling out the combination 6, 6, and 1, and leaving 9, 2 and 2 as the answer.

92. Four Families

The reason we figured we might have to skip the game was that, thanks to the flu, fewer than 18 kids showed up at the park that Saturday.

There are only fourteen different combinations of numbers that add up to 17 or less and still allow each of our four families to have a different number of kids. Eleven of the fourteen possible combinations give you different totals if you multiply the numbers together, the other three all give you the same total: 120.

That's how many votes I told the wiseguy he was going to get on election day; that's why he had to ask his question about the Blacks.

The three combinations that give you 120 when the numbers are multiplied are: 8, 5, 3, 1; 6, 5, 4, 1; and 5, 4, 3, 2. So when the would-be mayor asked me "Do the Blacks have more than one child?" I must have answered "Yes," otherwise he wouldn't have been able to figure out how many kids each of our families had: 5, 4, 3 and 2.

93. Father and Grandfather

Yes. It is perfectly possible for one's maternal grandfather to be younger than one's father.

94. Husbands and Fathers

After their respective parents divorced, or their mothers died, Mary and Joan married each other's father. Alternatively, of course, their respective parents may never have been married in the first place.

95. How Many Children?

Four boys and three girls.

96. Boys and Girls

The traditional answer is as follows:

No, the sultan's plan would not work. Of the first children born to all the women, half would be boys and half girls. Only the latter half would be allowed to bear a second round of children, and, again, half of this round would be boys and half girls. Again, the mothers of the boys would drop out, leaving (on average) a quarter of the original number of mothers to have a third round of children, which again would be evenly split between boys and girls.

In any round of births, the ratio between boys and girls would never change and, therefore, in aggregate there will always be as many boys as girls.

The above solution is not entirely accurate. In fact, the sultan's law *may indirectly* result in more girls being born than boys or, for that matter, more boys being born than girls. This is not for the reason the sultan had in mind, but because of the way the law of averages operates.

The question assumes that the natural birth ratio between boys and girls is 50–50. But like all averages, the average only holds true over an infinite number of attempts (in this case, births). Over any finite number, the average is unlikely to hold precisely true, and the smaller the finite number the greater the likely deviation from the average. Thus, in the first round of births in the sultan's country, the incidence of boys and girls is unlikely to be exactly equal; how one-sided it is will depend on how many expectant mothers there are. If the first round of births produces more girls than boys, the sultan's law will have the effect of reducing the opportunity for the law of averages to "even out" the discrepancy; this evening-out opportunity is reduced progressively as every round of births further restricts the pool of mothers. The same is true if the first round of births produces more boys than girls.

97. Two Children

Since I have two children, at least one of which is a boy, there are three equally probable cases: Boy–Boy, Boy–Girl or Girl–Boy. In only one case are both children boys, so the probability that both are boys is 1–3.

My sister's situation is different. Knowing that the older child is a girl, there are only two equally probable cases: Girl–Girl or Girl–Boy. Therefore the probability that both children are girls is 1–2.

98. Marital Problem

Jason and Dean were both clergymen. Dean married Jason to Denise, which explains why they share the same wedding anniversary. Jason married Jackie to a man whose name happens to be Peter. On another occasion, John married Dean to a girl called Paula.

99. The Severed Arm

The nine men, including John, were shipwrecked together on a remote, uninhabited island in the South Pacific during the Second World War.

Their only hope of rescue was a search operation by the US Air Force or Navy. As the days passed and their rations ran out, they faced a hard choice between dying one by one or cannibalizing parts of their bodies. They agreed on the latter course, beginning by sacrificing their left forearms. They drew straws to determine in what order they would have their limbs severed, but first they swore a pact that, should they be rescued before they had each had an arm removed, those in the group remaining whole would make arrangements to have their left arms amputated later.

The group was rescued after 8 of the 9 men had had an arm severed. The war ended shortly afterwards. Once back in the civilian world, John quickly thought better of his promise, and hit upon a scheme to deceive his comrades into believing he had honored the pact.

100. The Elevator Stopped

The woman's husband depends on a life support system connected to the electricity supply in their apartment on the tenth floor. What happened to the elevator made the woman realize the building was suffering a power failure.

101. The Barber Shop

David and the barber's wife are having an affair. They arrange their meetings for the end of the day, and David checks on the barber's workload to make sure he won't be home too soon.

102. The Car Crash

Shame on you, you male chauvinist! The surgeon was Mrs.—er, *Ms.*—Jones, Robert's mother.

103. A Glass of Water

The man had hiccups. The bartender's action, producing a sudden shock, was a quicker-acting cure than a glass of water might have been.

104. The Unfaithful Wife

On Eva's way out, John had noticed a run in her left stocking. When she went to the kitchen for coffee, he noticed that the run was on her right leg.

105. The Suicide

Plod wasted his last years. The husband was completely innocent. The heiress had committed suicide by tying a noose around her neck, standing on a block of ice and kicking the ice out from under her. It had fallen into the bathtub (or Jacuzzi) and melted overnight while her husband was away.

106. The Deadly Scotch

The poison was inside the ice cubes, which dissolved in Lay's drink, but not in El's.

107. The Heir

The test was a blood test. The elder remembered that the true prince was a haemophiliac.

108. Death in the Car

The victim was in a convertible. He was shot when the top was down.

109. A Soldier's Dream

The soldier had his dream while on guard duty.

110. The Antique Candelabrum

The elderly man had mentioned to the dealer that the candelabrum was one of a rare pair that together were worth much more than twice the value of the one; the man told the dealer he had been looking for the pair for many years. He bought the one candelabrum for $5,000, making it clear that he would pay handsomely if the dealer would locate its twin.

Not realizing that the elderly man was a confidence trickster, the dealer then called around some of his friends in the trade until he was tipped off about a collector who was offering to sell a candelabrum just like the one the dealer had sold to the elderly man.

Triumphantly, the dealer tracked down this "second" candelabrum to Robert, agreed to buy from him for $9,000, expecting to make a killing when he sold it on to the elderly man as the second part of the complete pair. Needless to say, the elderly man disappeared without trace but *with* $4,000 profit!

111. The Two Accountants

Fred Jones and Helen Smith were lovers, Smith having deceived Jones into believing she was single and interested in marriage.

112. The Judgment

The convicted killer was one of Siamese twins.

113. The Jilted Bride

No. She had simply jammed the clapper with packed snow, which melted during the course of the ceremony.

114. The Elevator Rider
Bill walks up eight floors, not by choice but because he is only four feet six inches tall, and cannot reach higher than the fifteenth floor on the elevator's panel of buttons. Whenever he has company on the ride up in the elevator, he asks someone to press 23 for him.

115. Crafty Cabby
The woman realized the cabby could hear because he drove her to her requested destination.

116. The Sharpshooter
The sharpshooter's hat was hanging over the barrel of his gun.

117. Death In Squaw Valley
The clerk remembered having issued flight tickets to the banker, who had booked a round trip for himself and a one way for his wife.

118. Burglars
Arthur is 10 months old.

119. Insomnia
Pete was being kept awake by Dave's snoring.

120. The North Pole
The odds are zero. In other words, the distance cannot be larger. For Eucla, or any other spot, to be 15,000 kilometers from the Pole it must be on the same parallel as Porto Allegre. But the circumference of the Earth at the Equator is near enough 40,000 kilometers and only slightly less (because of the flattening effect) around the Poles. Therefore the maximum distance between any two locations on the 30° parallel, via the South Pole, is approximately 10,000 kilometers.

121. The Long Division

10. Start at the end and work towards the beginning. Nine-tenths of 100 is 90. Eight-ninths of 90 is 80. Seven-eighths of 80 is 70. And so on until you come to one-half of 20.

122. Pool Resources

Jim has $5 and Andrew $7. There are two clues to the solution. The difference must be $2 to balance, and the original holdings must be odd numbers otherwise Andrew, having received $1, can never have twice Jim's amount.

Using an equation, let Jim's amount be x and Andrew's y.
Then:
$$2(x - 1) = (y + 1)$$
and
$$(y - 1) = (x + 1)$$
or
$$y = x + 2$$
Therefore:
$$2(x - 1) = x + 3$$
$$2x - 2 = x + 3$$
$$x = 5$$
$$y = 7$$

123. The Parking Dodge No. 1

After the meeting Clive approached the first policeman he could find and told him in broken German.

"Officer, I have a terrible problem, can you help me? I don't know my way around Zurich. I have parked my car somewhere near the lake and for the last hour I have been searching frantically, but I cannot find it."

The Swiss police might be tough on parking offenders but they are also helpful to foreign tourists. Clive's car was found within 30 minutes and, needless to say, there was no question of a penalty.

124. The Parking Dodge No. 2

He removed a parking ticket from another car and stuck it on his. On leaving the theater he replaced it, if the offending car was still there. If not he discarded it.

125. The Blip

The agent recorded a one and a half minute message on tape, then speeded it up two hundred times and transmitted it to Riyadh. The recipients slowed down the blip by the same factor.

They knew that it was their man who had sent the message because his speech pattern had been recorded on an oscilloscope which reduced it to a series of lines which are as positive an identification as a fingerprint. They also knew that he was transmitting as a free agent because they had agreed with him certain pre-arranged words or intonation to indicate if he were acting under duress.

126. The Crossroad

George, somewhat piqued, asked her:

"Aren't you going to drink?"

"Not until the police have been here," she replied.

127. A Wartime Mystery

When the KGB chief entered the conference room, Topolev recognized the civilian, Klaus Von Hasseldorf, aide-de-camp to the German ambassador Count Schulenberg. Topolev immediately realized that Hasseldorf was spying for the KGB. Topolev in turn was betraying his country to the Germans and in fact Hasseldorf had acted as his minder. Only a few days later on June 22nd 1941, at 4 A.M., Ribbentrop delivered a formal declaration of war to the Russian ambassador in Berlin.

128. The Black Forest

The man was a fighter pilot in the German air force. His plane had developed engine trouble and he had to bail out. His parachute got tangled up in the tree in such a way that the pilot was unable to free himself and after a few days he had died of exposure.

129. The Cabin in the Woods

The night before an explosion was heard and a fireball in the air noticed by several villagers, who reported it to the constabulary. It was assumed that the observation was caused by a plane crash and Hansel and Gretel were sent out to investigate. What they discovered in the woods was the cabin of an airplane.

130. The Sixpack

The answer is more complex than one would think. To start with, George would feel a reduction of the load he was carrying. His total weight (including remaining cans) would not diminish but the strain on his muscles would. After a short time George would lose weight through increased perspiration.

131. The Gamblers

Ian and Emma had booked a suite aboard a luxury cruise liner. On the fourth day there was an explosion in the engine room as a result of which the ship sank. Some passengers were saved though many passengers drowned, including Ian and Emma.

132. Spirit of St. Louis

Lindbergh was twice as safe with his single-engine plane.

The argument runs as follows:

Suppose the manufacturer has produced a batch of 100 engines of which one was faulty. The probability that this was Lindbergh's engine was 1%. Had he flown a twin-engined plane from the same batch the probability would have been 2%.

133. The Appointment

Gerry was broke and decided to hitchhike. He was unlucky in getting lifts and lost time waiting in Long Beach and Oceanside.

134. A Pair of Socks

Three. If I pick three socks, then either they are all of the same color (in which case I certainly have a pair of the same color) or else two are of one color and the third is of the other color, so in that case I would again have a matching pair.

135. Dubliners

The answer to the first question is yes. Assume there are exactly one million people living in Dublin. If each inhabitant had a different number of hairs, then there would be one million different positive whole numbers each less than one million—which is impossible.

The answer to the second problem is 450. Assuming that one islander is completely bald, there are 450 variants between 0 and 449 hairs; but, since we know that no islander has 450 hairs, from 451 hairs onwards the number of variants is the *same* as the maximum number of hairs, which means that the third fact stated in the question would not hold good for any number of inhabitants higher than 450.

136. The Clock-watcher

When George left his house, he started the clock and wrote down the time it then showed. When he got to his sister's house he noted the time when he arrived and the time when he left. He thus knew how long he was at his sister's house. When he got back home, he looked at the clock, so he knew how long he had been away from home. Subtracting from this the time he had spent at his sister's house, he knew how long the walk back and forth had been. By adding half of this to the time he left his sister's house, he then knew what time it really was now.

137. The Prisoners' Test

The man is wearing a red hat. His reasoning is as follows: "The first man did not see 2 white hats. If he had, he would have known immediately that he was wearing a red hat because there are only 2 white hats. The second man, aware that the first did not see 2 white hats, needed only to look at me; if he saw a white hat on me, he would know he was wearing a red hat (otherwise the first man would not have been stumped). Since he didn't know, he could not have seen a white hat on me. Therefore, my hat must be red."

138. Above or Below

The Z goes above the line. The pattern is so simple that many intelligent people miss it: letters consisting of straight lines go above, letters with curves go below!

139. Strange Symbols

Alice realized that, behind a looking glass, everything is reflected. The symbols stand for the numbers, one to seven. The right-hand side of each symbol is the correct numeral, the left-hand side is its mirror image. The next symbol in the sequence, then, is a back-to-back figure 8.

140. Product

Since one of the terms in this series will be $(x - x)$, which equals zero, the product of the entire series is zero.

141. What Are They?

House numbers.

142. Mending the Chain

The jeweler cut all 3 links on one of the pieces, then used the broken links to join the other segments. He charged $6.

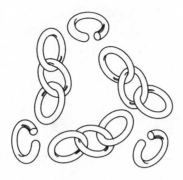

143. Fast Fly

At first glance it may seem that a horrendous calculation is necessary to solve this: the sum of an infinite series of numbers that get smaller and smaller as the cars approach each other. But if you focus on time rather than distance, a solution is easy. The cars are 50 miles apart and traveling towards each other at a combined speed of 50 miles per hour, so they will meet in one hour. In that hour, a fly that flies at 100 miles per hour will naturally travel 100 miles.

144. How Fast?

Speedy will have to drive at an infinite speed in order to average 100 miles per hour for the course. He must drive the whole 1,000 miles in 10 hours to attain the required speed, but he has already used up his 10 hours to drive the first half of the course. He will have to finish the race in zero time!

145. Which Coffeepot?

If these are typical coffeepots, they will both hold the same amount. A pot can be filled only to the level of its spout, otherwise the coffee will spill out. The spouts on both pots rise to the same height.

But these pots may not be typical. If they have hollow handles, as the drawing suggests, the smaller pot would hold more coffee because its entire handle is below the spout line.

146. Cocktail

You can do it by moving only two toothpicks. Slide the horizontal one over half a length, then bring down one of the vertical toothpicks to complete the upside-down glass.

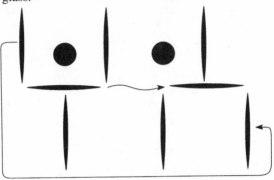

147. The Explorer and the Bear

The starting point could be anywhere on a circle drawn around the South Pole at a distance slightly more than $1 \times \frac{1}{2}\pi$ miles (about 1.16 miles) from the Pole. The distance has to be slightly more to take into account the curvature of the Earth. After you have walked a mile south, walking 1 mile east will take you on a complete circuit around the Pole. Finally, walking 1 mile north will return you to your starting point. Thus the starting point could be any one of the infinite number of points on the circle with a radius of about 1.16 miles from the South Pole. However, you could also start at points closer to the Pole, so that the walk east would take you exactly twice around the Pole, or three times, or four times, etc. (though, of course, if you started from any point closer to the Pole than a mile away, the direction of the initial stretch of one mile south would make you reach the Pole, after which you would actually be walking due north "up" the other side).

148. Buttons and Boxes

You can ascertain the contents of all three boxes by taking out just one button. The solution depends on the fact that the labels on all three boxes are incorrect.

Take a button from the box labeled RG. Assume that the button removed is red. You now know that the other button in this box must be red also, otherwise the label would be correct. Since you have identified which box contains 2 red buttons, you can work out immediately the contents of the box marked GG because you know it cannot contain 2 green buttons since its label has to be wrong. It cannot contain 2 red buttons, for you have already identified that box. Therefore it must contain one red and one green button. The third box, of course, must then be the one with the 2 green buttons. The same reasoning works if the first button you take from the RG box happens to be green instead of red.

149. Manhattan and Yonkers

Although the trains to Yonkers and Manhattan arrive equally often—every 10 minutes—it so happens that the Manhattan train always arrives one minute after the Yonkers train. Thus, the Manhattan train will be the first to arrive only if Sharon happens to arrive at the station during this 1-minute interval. If she enters the station at any other time, during the 9-minute interval, the Yonkers train will arrive first. Since Sharon's arrival is random, the odds are 9 to 1 in favor of Yonkers.

150. Counterfeit Coins

Only a single weighing is necessary to identify the counterfeit stack. Let x be the weight of a genuine silver dollar. Take 1 coin from stack No. 1, 2 from stack No. 2, 3 from stack No. 3, and so on to the entire 10 coins from stack No. 10. Weigh the whole sample. The sample should weigh $55x$. The number of grams over or under $55x$ the sample weighs corresponds to the number of the stack containing the counterfeit coins. For instance, if the sample weighs 7 grams more than it should (or 7 grams less), then the stack containing the counterfeit coins is No. 7.

151. Fake!

The point of this puzzle is that the counterfeit coin has to be identified in a limited number of weighings even though, at the outset, we do not know whether the counterfeit is heavier or lighter than a genuine coin.

The key to solving it lies in the fact that, as the first and second weighings narrow the field, we learn that certain coins can only be either genuine or light (but not heavy), and certain others can only be genuine or heavy.

In the following explanation it is assumed that, whenever a weighing reveals an imbalance, it is the left scale that is heavy. Of course, the right scale is just as likely to be heavier, but this does not affect the reasoning; it simply reverses the conclusions to be drawn about the coins in each scale. Coins proven to be genuine are designated by the symbol x.

1. Weigh coins 1, 2, 3, 4 against coins 5, 6, 7, 8.

   If:

   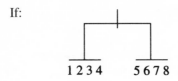

   1 2 3 4   5 6 7 8

2. Weigh coins 9, 10 against coins 11, x (x being any of coins 1–8).

   If:

   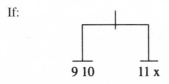

   9 10   11 x

   Coin 12 is the counterfeit. (A third weighing would determine whether it is heavy or light.)

   If:

   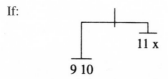

   9 10   11 x

   Then, obviously, the counterfeit is either 9, 10 or 11. In the example shown, we can tell that *either* 9 or 10 is heavy, *or* 11 is light. (Similarly, if the scale were reversed, we would know for certain that *either* 9 or 10 is light, *or* 11 is heavy.)

3. Weigh coins 10, 11 against x, x.
   If they balance, 9 is the counterfeit (and we know whether it is light or heavy depending on the result of the second weighing). If they do not balance, then 10 or 11 must be the counterfeit coin, and we can determine which by comparing the result of the second weighing with that of the third.

If:

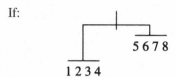

We know that coins 9–12 are genuine. Also that *either* one of 1, 2, 3 and 4 is heavy, *or* one of 5, 6, 7 and 8 is light.

4. Weigh coins 1, 5, X against 2, 6, 7.

If:

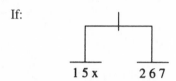

Then we know *either* 3 or 4 is heavy, *or* 8 is light.

5. Weigh coin 3 against 4.
   If they balance, coin 8 is the counterfeit and it is light. If they do not balance, whichever is the heavier is the counterfeit.

If:

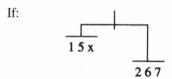

Then we know *either* 1 is heavy, *or* 6 or 7 is light.

6. Weigh coin 6 against 7.
   If they balance, coin 1 is the counterfeit and it is heavy. If they do not balance, the lighter is the counterfeit.

If:

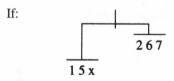

Then *either* 5 is light *or* 2 is heavy (the only explanations for the imbalance switching).

7. Weigh either coin 5 or coin 2 against x to determine which is the counterfeit coin.

152. How Long?

19.2 yards.

Let x be the length of the bolt. Then $\frac{x}{3} + \frac{x}{4} + 8 = x$. This reduces to:

$$4x + 3x + 96 = 12x, \text{ or } 5x = 96$$

$\therefore$ $x = 19.2$ yards.

153. Arithmetic Problem

Ten problems.

Let x be the number of correct solutions, and y be the number of incorrect solutions.

$$
\begin{aligned}
\text{Then:} \qquad x + y &= 26 \\
\text{and} \qquad 8x - 5y &= 0 \\
\text{From first equation } y &= (26 - x) \\
\therefore \qquad 8x - 5(26 - x) &= 0 \\
\text{or} \qquad 8x &= 130 - 5x \\
\therefore \qquad x &= 10
\end{aligned}
$$

154. The Cloak

x = value of the coat. After seven months the butler is entitled to $\frac{7x}{12} + \frac{700}{12}$. However, he receives only \$20, therefore, $\frac{5x}{12}$ must compensate for the difference between $\frac{\$700}{12}$ and $\frac{\$240}{12}$ (\$20)

therefore, $5x = 460$

and $x = \$92$

155. Water Lentils

As in fact you are starting with two lentils on the second day, you save one day, therefore the answer is 29 days.

156. Two Steamers

The second, because

$$\frac{x}{30} + \frac{x}{40}$$

is greater than

$$\frac{2x}{35}$$

Proof:

$$\frac{x}{30} + \frac{x}{40} = \frac{7x}{120} \text{ or } \frac{245x}{4200}$$

$$\frac{2x}{35} = \frac{240x}{4200}$$

157. Unequal Scales

It depends on the reason for the scales' imbalance. If the pans are of unequal weights, the grocer's solution will work; but if the arms of the scale are of unequal lengths, it will not, and the grocer will lose. This can be shown algebraically as follows:

Let a and b be the unequal lengths of the arms of the scale, x the fixed weight, and y and z the amounts of, say, sugar weighed out. Then $ax = by$ and $bx = az$, so the weight of sugar dispensed is:

$$\left\{ \frac{a}{b} + \frac{b}{a} \right\}$$

which is greater than $2x$. For

$$\frac{a}{b} + \frac{b}{a} > 2; \ a^2 - 2ab + b^2 > 0; \text{ and } (a - b)^2 > 0$$

are equivalent statements when a and b are distinct positive quantities.

It can also be explained in terms of physics. While adding a compensating weight to a light pan will bring the scales permanently into balance, adding a compensating weight to the side with a short arm will only put the scales into momentary balance. The combined effect of a long arm and weight added to that side (in the form of goods, for instance) will be to increase the imbalance progressively, so that ever more compensation is needed.

158. The Long Division

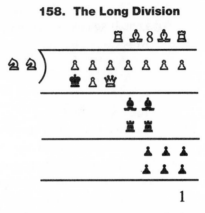

First, observe that the five-digit quotient forms only three products with the divisor. Therefore, two of the five digits must be zeros. These cannot be the first or last, since both obviously form products. They are therefore the second and fourth digits, those covered by the white Bishops. Furthermore, the two-digit divisor, when multiplied by 8, gives a two-digit product; but when multiplied by another number, the one concealed under the first white Rook, it gives a three-digit product; the multiplier hidden under the first white Rook must therefore be larger than 8, namely 9. Both the first and last digits in the quotient give three-digit products with the two-digit divisor; both must therefore be 9. We now have established the quotient: it is 90,809. Let us find the divisor, covered up beneath the two white Knights. When multiplied by 8 it forms a two-digit product; when multiplied by 9 it forms a three-digit product. It must, therefore, be 12; $8 \times 12 = 96$, $9 \times 12 = 108$; neither 10, 11, 13 nor any larger number meets these requirements. The numbers under the remaining chess pieces are now readily ascertainable.

159. Two Bolts

The bolt heads will remain at the same distance from each other in both cases.

Assume that bolt A has not a thread but parallel rings and you swing bolt B around it, the heads will approach or move away, depending on whether you swing counter- or clockwise. As, however, bolt A has the same thread as B they cancel each other out.

Another explanation:

If instead of swinging bolt B around bolt A, you leave B in situ but turn it clockwise, then the heads will approach. Indeed this is the function of a screw, as bolt "A" can be considered a nut. If you turn B counterclockwise, you "unscrew" and the heads will move away. However if you do neither but just swing B around then the heads will remain where they are.

160. The Duel

The poorest shot, the Baron of Rockall, has the best chance of surviving. Lord Montcrief, the one who never misses, has the second best chance. Because the baron's two opponents will aim at each other when their turns come, his best strategy is to fire into the air until one of the others is killed. He will then get the first shot at the survivor, which gives him the advantage.

161. Fair Shares

There are several possible solutions. However, the following method has the advantage of leaving no excess pieces of cake.

Assume 4 people are sharing the cake. Call them A, B, C and D. First, A cuts off what he is content to keep as his $\frac{1}{4}$ of the cake. Next B has the option, if he thinks A's slice is more than $\frac{1}{4}$, of reducing it by cutting off some of it. If B thinks A's slice is $\frac{1}{4}$ or less, he does nothing. C and D in turn then have an opportunity to do the same with A's slice. The last person to touch this slice keeps it as his share. If anyone thinks that this last person has less than $\frac{1}{4}$ he is naturally pleased because it means, in his eyes, that more than $\frac{3}{8}$ remains. The remainder of the cake, including any cut-off pieces, is now divided among the remaining 3 persons in the same manner, then among 2. The final division is made by one person doing the cutting and the other the choosing. This procedure can be applied to any number of persons.

162. Racing Driver

160 miles per hour. The key lies in converting miles per hour to miles per minute and in using fractions instead of decimals to avoid rounding errors.

Using the formula Time (T) = $\dfrac{\text{Distance (D)}}{\text{Speed (S)}}$.

If driver travels 3 miles at 140 miles per hour, it takes him $\dfrac{9}{7}$ minutes to cover the distance.

Equally, $1\frac{1}{2}$ miles at 168 miles per hour will take $\dfrac{15}{28}$ minutes and $1\frac{1}{2}$ miles at 210 miles per hour takes $\dfrac{3}{7}$ minutes.

∴ Total time for six miles = $\dfrac{9}{7} + \dfrac{15}{28} + \dfrac{3}{7}$ minutes.

This reduces to $\dfrac{63}{28}$ minutes for 6 miles.

Using above formula S = $\dfrac{D}{T} = \dfrac{6 \times 28}{63}$ per minute

or: $\dfrac{6 \times 28 \times 60}{63}$ per hour

This reduces to 160 miles per hour.

163. The Side View

Side view

Perspective

164. The Striking Clock

Four seconds—the time between the clapper striking the bell for the first peal and the second one is 2 seconds, and 2 seconds later it strikes for the third peal. Do not be confused by the lingering sound—I said strike!

165. Red or Green

Most people erroneously include No. 4 in their answer. But consider: No. 2 does not matter, since the question is concerned only with red cards. If No. 1 has a circle, the answer to the question is NO. Similarly, if No. 3 is red the answer is NO. If No. 1 is a square, No. 3 is green, and No. 4 is either red or green the answer is YES. Therefore the answer is: No. 1 and No. 3.

166. The River

$5\frac{1}{3}$ (5 minutes 20 seconds)

Speed downstream is $\frac{1}{2}$ km. per minute. Return speed is $\frac{1}{4}$ kilometer per minute. Therefore the current makes $\frac{1}{8}$ of a kilometer difference per minute.

Consequently, the boat speed is $\frac{3}{8}$ of a kilometer per minute, which translates into $5\frac{1}{3}$ minutes for the 2 kilometers in still water.

167. Changing the Odds

The boy picked a pebble out of the hat and, before they had a chance to examine it, dropped it, apparently accidentally, where it was lost among the pebbles on the ground. He then pointed out to the king that the color of the dropped pebble could be ascertained by checking the color of the one remaining in the hat.

168. Zeno's Paradox

Achilles would reach the tortoise at $1,111\frac{1}{9}$ meters. If the race track is shorter than this, the tortoise would win. If it were exactly this size, it would be a tie. Otherwise Achilles will pass the tortoise.

169. The Boy and the Girl

The boy has red hair, the girl black hair. There are four possible combinations: true-true, true-false, false-true, and false-false. It is not the first, since we are told that at least one statement is false. Nor is it the second or third because, in each case, if one lied, then the other could not have been telling the truth. Therefore it is the fourth; both lied.

170. Infinity and Limits

The increasing radii will in fact approach a limit that is about 12 times that of the original circle.

171. Hats In the Wind

The probability is zero. If 9 people have their own hats, then the tenth must too.

172. Handshakes

Even. If you were to ask everyone in the world how many hands he or she has shaken, the total would be even because each handshake would have been counted twice—once each by the two people who shook hands. A group of numbers whose sum is even cannot contain an odd number of odd numbers.

173. Move One

Move the vertical bar of the + sign to the other side of the equation so that it now reads:

$$\text{V I} - \text{II} = \text{IV}$$

There is a second solution:

$$\text{V II} - \text{II} = \text{V}$$

174. A Chiming Clock

An hour and a half—from 12:15 to 1:45. When you have heard the clock chime once 7 times, you need not wait for it to chime again, for the next cannot be anything but 2 o'clock.

175. Ring Around the Circle

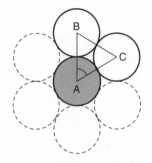

Exactly 6 circles will complete the ring. For the proof, inscribe the ∠ ABC in the center circle and two of the outer circles. The triangle is equilateral, since each side is equivalent to two radii. In equilateral triangles, each angle is 60°. Therefore ∠ A is 60°, or one-sixth of a 360° circle.

176. A Unique Number

It is the only one that contains all the numerals in alphabetical order.

177. The Boss and His Chauffeur

For 50 minutes. He saved the chauffeur 10 minutes of traveling time each way and thus was picked up at 4:50 P.M., rather than the usual time.

178. Weather Report

The five temperatures were: 1, –1, 2, –2, and 3.

179. Can You?

This statement was made by a salmon fisherman who was asked what he did with all the fish he caught.

180. Presidents

If the fifth president were not among those who died on that date, then the newspaper item would almost certainly have made the more impressive statement: "Three of the first four presidents died on the Fourth of July." Therefore, Professor Flugel was reasonably confident that the fifth president, James Monroe, died on that date.

181. Jim and George

They are standing back-to-back.

182. Two Sizes of Apples

This is a common trap in mathematical tests. The charge for the apples should be $33\frac{1}{3}$ cents for large apples and 20 cents for smaller apples, so the average charge per apple should be:

$$(33\tfrac{1}{3} + 50)/2 = 26\tfrac{2}{3} \text{ cents}$$

and not 25 cents, which the boy collected. If the 60 apples had been sold for $26\frac{2}{3}$ cents each, the boy would have received

$$60 \times 26\tfrac{2}{3} \text{ cents} = \$16$$

The son was charging too little for the apples and the dollar went to the customers.

183. The Sequence

The solution is:

$$O \;\; T \;\; T \;\; F \;\; F \;\; S \;\; S \;\; E \;\; N \ldots$$

being the initial letters of One, Two, Three, Four, Five, Six, etc.

184. The Half-full Barrel

All they had to do was tilt the barrel on its bottom rim till the water was just about to pour out. If the barrel is exactly half full, the water level at the bottom of the barrel should just cover all the rim. That way half the barrel is full of water; the other half is air space. If the water amply covers the bottom rim, the barrel is more than half full; if the bottom is not fully covered, the barrel is less than half full.

185. The Marksman

The marksman who fires 5 shots in 5 seconds takes $1\frac{1}{4}$ seconds between shots, since there are 4 intervals between the first and last shots. The other marksman requires 10 seconds for 9 intervals, or $1\frac{1}{9}$ seconds between shots. Therefore, the second marksman will take less time to fire 12 shots—$12\frac{2}{9}$ seconds compared with $13\frac{3}{4}$ seconds.

186. The Square Table

Let's examine the problem from the point of view of any one of the women. Initially, since she is a woman, she reasons that her neighbors' hands would be raised regardless of whether the unseen individual was a man or a woman. But after further thought it occurs to her that if the person opposite was actually a man, he would have known immediately that the person sitting opposite him was a woman because his neighbors would have raised their hands only for that reason. Since the unseen person did not make this announcement, it could only be because she was a woman.

187. The Marbles

Only 2 marbles can be transferred out of the first bag. The contents of the 2 bags will then be one of the following:

	First bag			Second bag		
	Col. A	Col. B	Col. C	Col. A	Col. B	Col. C
1st possibility	3	3	1	3*	3*	5
2nd possibility	3	2	2	3	4	4

To assure at least 2 of each color in the first bag, at least 7 marbles must be transferred back, because the first 6 might be the 3 color A and the 3 color B marbles represented by the starred 3s in the first possibility shown above. Therefore, there will be 4 marbles remaining in the second bag.

188. A Carbon Copy

He must start in the lower left corner.

189. Walking in Step

Whenever they step out together, it will always be on the left foot.

190. Razor Shortage

One's first thought is that the first shopper would take $3 and the second $5. But this is not correct.

The $8 was in payment for $\frac{8}{3}$ packs of razors. It follows then, that the equivalent of 8 full packs would be $24. So one pack is worth $3. Since we each ended up with $\frac{8}{3}$ packs, the first shopper, who had 3 packs to start with, gave me $\frac{1}{3}$ of a pack; the other $\frac{7}{3}$ were given by the other shopper. Therefore, $1 goes to the first shopper and $7 to the second.

191. Unusual Equations

(a) $\overset{\downarrow}{5}45 + 5 = 550$

(b) $99 + \frac{9}{9} = 100$

(c) $.\overline{7} \times .\overline{7} = 100$

(This solution is somewhat flawed, as, strictly speaking, zeros should be used thus: 0.7. Besides, the same principle can be applied to any number; n/0.n × n/0.n always equals 100!)

(d) $\dfrac{9+9}{.9}$

(Again, the divisor should read 0.9.)

192. The d'Alembert Paradox

No, his reasoning was incorrect. D'Alembert made the error of not carrying through his analysis far enough. The three cases are not equally likely, and the only way to obtain equally likely cases is, in the third case, to toss the coin again even when the first toss is heads; so that the third case has, in fact, two options and becomes the third and fourth cases. The four possible cases are, therefore, as follows:

(a) Tails appears on the first toss and again on the second toss.

(b) Tails appears on the first toss and heads on the second toss.

(c) Heads appears on the first toss and again on the second toss.

(d) Heads appears on the first toss and tails on the second toss.

As there are now proved to be four cases and as three of these are favorable, then the probability of heads at least once is, in fact, 3/4.

193. Jugs

To solve this puzzle you must first investigate the only two possibilities by which you can begin the decanting: you can either pour water into jug B until jug B is full, or pour water into jug C until jug C is full. During the operations you must avoid a situation in which both B and C are entirely full because then the only way to proceed would be to pour the contents of B and C entirely into A—in other words go back to the beginning and start again.

The two possibilities are:

JUG	A	B	C	A	B	C
Commence (pints)	8	0	0	8	0	0
Operation 1	3	5	0	5	0	3
Operation 2	3	2	3	5	3	0
Operation 3	6	2	0	2	3	3
Operation 4	6	0	2	2	5	1
Operation 5	1	5	2	7	0	1
Operation 6	1	4	3	7	1	0
Operation 7	4	4	0	4	1	3
Operation 8				4	4	0

Thus it can be seen that to commence by pouring into jug B until it is full produces the solution with the least number of decantings, which is 7.

194. One-two-three

Each line of numbers describes the line above it, i.e. 1, then 1 (one) 1, then 2 (two) 1s, then 1-2, 1-1 etc. The next row is 3 1 1 3 1 2 1 1 1 3 1 2 2 1.

195. A Bottle of Wine

Hands up everyone who said $1. This is wrong, as the total of the bottle of wine will then be $11. The correct answer is 50 cents.

196. Roll-a-penny

To win, the punter's coin must fall with the center within the shaded area (see figure below). If the center is outside the shaded area then the coin must touch the line somewhere and is a loser.

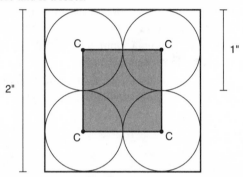

So winner area = 1 square inch; losing area = 3 square inches. Therefore odds should be 3 to 1 and should pay out 3 times the stake plus the stake (in the case of a $.02 stake, $.08). The actual payout is $\frac{80}{16}$ = $.05.

Therefore the odds favor the banker by 8 to 5.

197. The Missing £

The way the question is posed includes a piece of mathematical sleight of hand. It suggests that each man spent £9 plus £2 tip between them, but the £27 included the tip.

We should say:

The meal cost:	£25
The waiter's tip	2
Change	3
	£30

198. Gold Card

It certainly is not! It is 2–1 on that the gambler will win. In other words, he will win 2 games out of 3.

We are not dealing with cards here but with sides. There were 6 sides to begin with, 3 of each:

GOLD	SILVER
1	
1	
	1
	1
$\frac{1}{3}$	$\frac{1}{3}$

The card on the table cannot be the silver/silver card so that variant can be eliminated. We are left with:

GOLD	SILVER
1	
1	
$\frac{1}{3}$	$\frac{1}{1}$

We can see one gold side so we are left with:

GOLD	SILVER
1	
$\frac{1}{2}$	$\frac{1}{1}$

The reverse side can be GOLD, or GOLD, or SILVER. Odds 2–1 *on*.

199. The Airplane

Since the wind boosts the plane's speed from A to B and retards it from B to A, one is tempted to suppose that these forces balance each other so that total travel time for the combined flights will remain the same. This is not the case, because the time during which the plane's speed is boosted is shorter than the time during which it is retarded, so the overall effect is one of retardation. The total time in a wind of constant speed and direction, regardless of the speed or direction, is always greater than if there were no wind.

200. Skiing the Atlantic

Suppose that three boats are needed, so that two boats can transfer their fuel at the right moment to the third. Call these A, B, and C. All three start at once from Long Beach with full tanks. When they get one-eighth of the way around, they have used up a quarter of their fuel. C then divides its fuel into three equal parts; having three-quarters of its fuel left, it transfers a quarter to A, a quarter to B, and uses the remaining quarter to return to home port. (Notice that all the fuel in its tank is used up.)

A and B now have full tanks again. They speed on till they reach a quarter of the way around. Both then have three-quarters of their fuel left. B then transfers a quarter of its fuel to A, because the remaining half is needed so that it can return to base. A full tank, we know, is sufficient to cross half-way. Since A has covered a quarter of the way with a full tank, it can cover three-quarters of the way. Here it is met by C, which, in the meantime, has refueled and sped from Long Beach in the other direction, using the fact that the Earth is round. It halves the remainder, so that both reach seven-eighths of the way, where they are met by B. To reach this point, B has used up a quarter of its fuel. It needs another quarter for the return journey, but divides the remainder between A and C. Now A, B and C can all return to base, A having successfully traveled around the Earth.

201. Four Bugs

10 inches. Since the paths are always perpendicular to each other, the original distance remains unchanged.

202. Birth Dates

With 24 people in the room you would, in the long run, lose 23 and win 27 out of each 50 bets.

203. Changing Money

It is quite obvious that the economies of Eastland and Westland paid for the razor blades. If Malcolm were to repeat the transaction often enough, the end result would give him all, or a large part, of the stock of razor blades in both countries, together with one Eastland or Westland dollar. The two countries would be left with their stocks of blades largely denuded, but with their domestic currencies repatriated.

204. Rice Paper

The stack will clearly consist of 2^{50} sheets of paper, which is well over 17 million miles high.

205. Two Discs

It is generally argued that since the circumferences are equal, and since the circumference of A is laid out once along that of B, A must make one revolution about its own center. But if the experiment is tried with, say, two coins of the same size, it will be found that A makes two revolutions. This fact can be shown diagrammatically as follows:

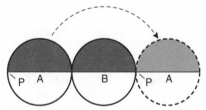

Let P be the extreme left-hand point of A when A is in its original position. A moment's thought will make it clear that when A has completed half its circuit about B, the arc of the shaded portion of A will have been laid out along that of the shaded portion of B, and P will again be the extreme left-hand point of A. Hence A must have made one revolution about its own center. The same argument holds for the arcs of the unshaded portions of A and B when A has completed the second half of its circuit about B.

206. The Slab and the Rollers

Suppose we resolve the motion into two parts. First think of the rollers lifted off the ground and supported at their centers. Then, if the centers remain stationary, one revolution of the rollers will move the slab forward one foot. Next, think of the rollers on the ground and without the slab. Then one revolution will carry the centers of the rollers forward one foot. If now we combine these two motions, it becomes clear that one revolution of the rollers will carry the slab forward a distance of 2 feet.

207. The Broken Stick
1 chance in 4.

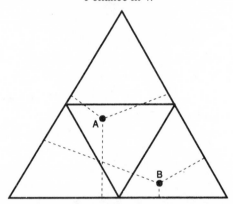

If we take any point in the large triangle, the point must fall within the smaller triangle for the 3 pieces to form a triangle. If it falls outside as point B, one side will be longer than the sum of the other 2 sides, and will not form a triangle.

208. The Slow Horses
"Change horses."

209. Hourglass
Start both hourglasses. When the 4-minute glass runs out, turn it over (4 minutes elapsed). When the 7-minute glass runs out, turn it over (7 minutes elapsed). When the 4-minute glass runs out this time (8 minutes elapsed), the 7-minute glass has been running for 1 minute. Turn it over once again. When it stops, 9 minutes have elapsed.

210. Sorting the Numbers

The numbers 15, 16 and 17 should be placed in groups 3, 3 and 2, respectively. Group 1 consists of numbers composed entirely of curved lines, Group 2 consists of numbers composed entirely of straight lines, and Group 3 consists of numbers composed of a combination of curved and straight lines.

211. Categories

The categories are as follows:

A M T U V W Y	(symmetry about vertical axis)
B C D E K	(symmetry about horizontal axis)
F G J L N P Q R S Z	(no symmetry) and
H I O X	(symmetry about both axes).

212. What Weights?

1 lb, 3 lb, 9 lb.

The key to this is that the boy can put any combination of weights on either pan and the difference between the two weights is the amount of fruit he sells. Thus the weight he requires is the result of an addition or subtraction sum.

$$1 - 0 = 1$$
$$3 - 1 = 2$$
$$3 - 0 = 3$$
$$(3 + 1) - 0 = 4$$
$$9 - (3 + 1) = 5$$
$$9 - 3 = 6$$
$$(9 + 1) - 3 = 7$$
$$9 - 1 = 8$$
$$9 - 0 = 9$$
$$(9 + 1) - 0 = 10$$
$$(9 + 3) - 1 = 11$$
$$(9 + 3) - 0 = 12 \text{ and finally}$$
$$(9 + 3 + 1) - 0 = 13.$$

213. Rope Trick

Half the rope (75 feet) plus the distance from the ground (25 feet) equal the height of the flagstaffs, thus the flagstaffs are right next to each other.

214. Catching the Bus

Juliette $16\frac{2}{3}$, or 30 meters, nearer Montreux. Lucille misses the bus everywhere.

Draw your own diagram of a straight road and letter the position of the bus B, the point where the sisters left it P, the patch of narcissi where Juliette is J, and the correct meeting point with the bus for Juliette M.

Let us call the distance PM x meters.

Juliette runs MJ meters during the time the bus travels BM meters.

$$\therefore\ 2\,MJ = 1\,BM$$

But $MJ = \sqrt{40^2 + x^2}$ (Pythagoras)

and $BM = 70 + x$

$$\therefore \qquad\qquad 2\sqrt{40^2 + x^2} = 70 + x$$

$$\therefore \qquad\qquad 6{,}400 + 4x^2 = 4{,}900 + 140x + x^2$$
$$\text{(after squaring each side)}$$

$$\therefore \qquad\qquad 3x^2 - 140x + 1{,}500 = 0$$

This reduces to: $\qquad (3x - 50).(x - 30) = 0$

For a product to be zero, one of its factors has to be zero

$\therefore$ Case (1) $\qquad\qquad 3x - 50 = 0\ $ or $\ x = 16\frac{2}{3}$

Case (2) $\qquad\qquad x - 30 = 0\ $ or $\ x = 30$

Thus Juliette could have run to a point nearer Montreux by either $16\frac{2}{3}$ or 30 meters from the point where they left the road and she would have caught the bus.

Lucille was 41 meters from the road, therefore:

$$2\sqrt{41^2 + x^2} = 70 + x$$

which becomes:

$$4(41^2 + x^2) = 4900 + 140x + x^2$$

This reduces to:

$$3x^2 - 140x + 1824 = 0$$

This equation will not give real roots and therefore Lucille will miss the bus.

215. Bus Timetable

The buses are evenly spaced along the road in both directions. The man notices buses at the rate of 30 an hour. Because he is moving towards one "stream" of buses and away from the other, he sees more buses in one direction than the other (20 to 10), but if he were stationary he would see 15 an hour traveling each way. The buses therefore leave the terminal at 4-minute intervals.

216. How Many Hops?

You will not be able to hop out.

You hop $4\frac{1}{2}$ feet at the first attempt, which is half-way out, and then another $2\frac{1}{4}$ feet at the next hop. Thus you are already three-quarters of the way out in two hops. You feel encouraged, for surely the last quarter will be hopped easily! Let us write down the hops:

$4\frac{1}{2}$, $2\frac{1}{4}$, $1\frac{1}{8}$, $\frac{9}{16}$, $\frac{9}{32}$, $\frac{9}{64}$, $\frac{9}{128}$, and so on.

Add these up and you will see that you are nearly there—in fact, you can hop more than $8\frac{3}{4}$ feet of the total distance needed of 9 feet. But this is a series whose "sum to infinity" is less than 9 feet. You are a prisoner in the circle!

217. Decaffeiné

$33\frac{1}{3}$ cups. Because there is 3 percent caffeine left in the doctored coffee; in 100 cups there would be enough for 3 cups of regular; 3 goes into 100 exactly $33\frac{1}{3}$ times.

218. Speed Test

Let 1234567891 be n. Then the denominator can be written as:

$$n^2 - [(n - 1) \times (n + 1)] \text{ or } n^2 - n^2 + 1, \text{ which} = 1$$

Therefore the answer is 1234567890.

219. Walking Home

Yes. He took as much time for the second half of his trip as the whole trip would have taken on foot. So no matter how fast the train was, he lost exactly as much time as he spent on the train.

He would have saved $\frac{1}{30}$ of the time taken by walking all the way.

220. The Watchmaker

As the problem says, the apprentice mixed up the clock hands so that the minute hand was short and the hour hand long.

The first time the apprentice returned to the client was about 2 hours and 10 minutes after he had set the clock at six. The long hand moved only from 12 to a little past 2. The short hand made 2 full circles and an additional 10 minutes. Thus the clock showed the correct time.

Next day around 7:05 A.M. he came a second time, 13 hours and 5 minutes after he had set the clock for six. The long hand, acting as hour hand, covered 13 hours to reach 1. The short hand made 13 full circles and 5 minutes, reaching 7. So the clock showed the correct time again.

221. The Lead Plate

They poured the shot into the jug and then poured in water, which filled all the spaces between the pellets. Now the water volume plus the shot volume equaled the jar's volume.

Removing the shot from the jar, they measured the volume of water remaining, and subtracted it from the volume of the jar.

222. The Caliper

He placed an object (such as a strip of wood) over one end of the cylinder, then rested one leg of the caliper against that object and the other leg inside the opposite indentation. The caliper could then be removed without opening the legs. He subtracted the thickness of the object from the spread of the caliper. This gave him a measurement equal to the length of the cylinder less one indentation. Subtracting this figure from the overall length of the cylinder gave him the depth of *one* indentation; doubling this gave him the depth of both indentations, which he could then deduct from the overall length of the cylinder.

223. Fuel Tanks

Since Pete Brown takes twice as much diesel as Joe Smith, the quantity of diesel must be divisible by 3. We know that we can divide numbers by 3 only if the sum of their digits is also divisible by 3. The sum of the digits on the storage tanks gives 6, 4, 1, 2, 7, 9. The sum of all these digits is 29 (which, when divided by 3, gives the same remainder as when 11 or 2 is divided by 3).

The capacity of the tank holding the special blend of unleaded and alcohol, when subtracted from the sum of the capacities of the other barrels, should leave a number which can be divided by 3. Therefore, if the capacity of the special blend tank is divided by 3, the remainder must be 2. If we look at the capacity of the tanks, we see that only the 20-gallon tank is the right size (sum of the digits is 2); 29 − 2 = 27, which is divisible by 3. Therefore, the unleaded/alcohol blend was contained in the 20-gallon tank.

This left 99 gallons of diesel, to be divided into 33- and 66-gallon consignments for Smith and Brown respectively. Thus, Smith was given all the fuel in the 15- and 18-gallon tanks, and Brown the fuel in the 15-, 19-, and 31-gallon tanks.

224. The Wire's Diameter

Wind a number of coils tightly around a cylinder as shown on the overleaf. Twenty diameters make 2 centimeters, so one diameter is 0.1 centimeters.

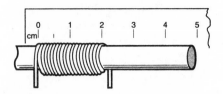

225. The Bottle's Volume

The area of circle, square or rectangle can easily be calculated after measuring sides or diameter with a rule. Call the area s.

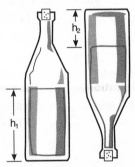

With the bottle upright (see illustration), measure the height h_1 of the liquid. The full part of the bottle has the volume sh_1.

Turn the bottle upside down and measure the height h_2 of the air space. The empty part of the bottle has the volume sh_2. The whole bottle has the volume $s(h_1 + h_2)$.

226. The Ship and the Seaplane

Perhaps you can spot without any algebra or extended calculation that the seaplane goes 200 miles, while the ship goes another 20.

This is a trial-and-error exercise. Now try to solve it by equation.

Let the distance be: X
The speed of the ship: S
The speed of the seaplane   10.S,
We also know that time = $\dfrac{\text{distance}}{\text{speed}}$ = $\dfrac{X}{S}$, then

$$\frac{X - 180}{S} = \frac{X}{10.S} \quad \text{(as time until they meet is the same)}$$

This reduces to:

$$10.\$.x - 1800\$ = X.\$$$
$$9X = 1800$$
$$X = 200$$

227. The Ships and the Lifebuoy

From the buoy's point of view (floating downstream), the ships move away from it at equal speeds in still water. Then they return at equal speeds in still water. Thus the two ships reach the buoy simultaneously.

228. Equation to Solve in Your Head

Adding and subtracting the equations we see that the numbers become 10,000, 10,000 and 50,000; and 3,502, –3,502 and 3,502. Dividing by 10,000 and by 3,502 we obtain:

$$x + y = 5$$

$$x - y = 1$$

Therefore: $x = 3$

$$y = 2$$

229. Three Men in the Street

The key is that the man in white is talking to Mr. Black and so cannot be he. Nor can he be Mr. White, since nobody is wearing his own color. So the man in white must be Mr. Gray. We can show what we know like this:

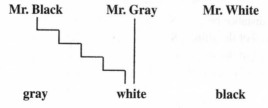

Mr. Black	Mr. Gray	Mr. White
gray	white	black

The straight line shows what must be true; the wiggly line shows what cannot be true. Mr. White cannot be wearing white; so he's in black. That leaves Mr. Black wearing gray.

230. The Square Field

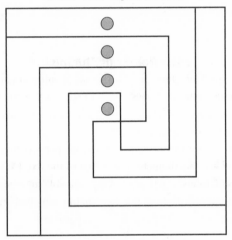

231. Kings and Queens

There are only two arrangements of Kings and Queens which can satisfy the first and second statements, these being KQQ and QKQ. The third and fourth statements are met by only two possible arrangements of Hearts and Spades, these being SSH and SHS. These two sets can be combined in four possible ways as follows:

$$KS, QS, QH$$
$$KS, QH, QS$$
$$QS, KS, QH$$
$$QS, KH, QS$$

The final set is ruled out because it contains two Queens of Spades. Since all the other sets consist of the King of Spades, Queen of Spades and Queen of Hearts, these must be the three cards on the table. It is not possible to state definitely which position any particular card is in, but the first must be a Spade and the third a Queen.

232. Even Tread

Each tire was used for four-fifths of the total mileage: four-fifths of 10,000 miles = 8,000 miles per tire.

233. Round and Round

Wheel A makes 3 revolutions about its own axis in rolling once around Wheel B. Since the circumference of Wheel A is half that of Wheel B, this produces two rotations with respect to Wheel B, and the revolution adds a third rotation with respect to an observer from above.

The general formula for the number of rotations per revolution is $(B/A) + 1$. So, if rolling Wheel A had a diameter twice that of the fixed Wheel B, it would rotate one and a half times. As it gets larger, the rolling wheel approaches a limit of one rotation per revolution, this limit being achieved only when it rolls around a degenerate "circle" (or point) of zero diameter.

234. Choose a Glass

The binary procedure is the most efficient method for testing any number of glasses of liquid in order to identify a single glass containing poison. First the glasses are divided as nearly in half as possible. Then one set is tested by taking a sample from each glass, combining them, and testing the mixture. The set identified as including the poisoned glass is then divided again as nearly in half as possible, and the procedure repeated until the poisoned glass is identified. If the number of glasses is between 100 and 128 inclusive, as many as 7 tests might be required. From 129 to 200 glasses might take 8 tests. The number 128 is the turning point. Since we know that the number was between 100 and 200 there must therefore have been 129 glasses in the hotel lounge, because only in that case would the initial testing of one glass make no difference in applying the most efficient testing procedure. To test 129 glasses by halving could result in 8 tests. If a single glass were tested first the remaining 128 glasses would require no more than 7 tests, so that the total number of tests remains the same.

When the above answer was first published, many people wrote to say that the detective inspector was right, and the statistician wrong. Regardless of the number of glasses, the most efficient testing procedure is to divide them as nearly in half as possible at each step and test the glasses in either set. When the probabilities are worked out, the expected number of tests of 129 glasses, if the halving procedure is followed, is 7.0155+. But if a single glass is tested first, the expected number is 7.9457+. This is an increase of 0.930+ test, so the inspector was almost right in considering the statistician to be wasting one test.

Put in simple terms, the statistician's suggestion would result in a wasted test in every case other than if the suspect glass happened to be the last one—the 129th—tested, and the chance of that occurring at random is not high.

235. The Square Window
See the illustration. The shaded area is the part of the window that is painted blue.

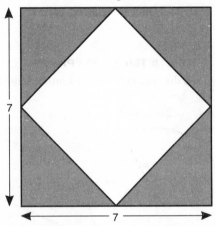

236. Dominoes
You can solve this with pages of diagrams and calculations—or, by an insight-ful shortcut, in just a couple of sentences. Every domino must cover 2 adjacent squares—that is, one black square and one white square. The diagonally opposite squares on a checker board are of the same color, both white in our example. You can arrange 30 dominoes so that they cover all 30 white squares and 30 of the black, but there will always be 2 black squares left, and the one remaining domino can't cover them both.

237. A Bridge Game

FLUSH: the two events are equally likely. You may prove this by doing pages of calculations or by using shortcut reasoning. If two players hold all the cards in one suit, the other two players are necessarily void in that suit—the two events occur together, hence they are equally probable.

PAPER PERFECT: All stories of perfect deals in bridge should be taken with a large pinch of salt. The odds against one are 2,235,197,406,985,633,368,301,599,999 to 1. This is so remote that a perfect deal has probably never occurred by chance (as opposed to by dint of a prank or a poorly shuffled deck) in the entire history of the game. If everyone in the world were dealt 60 bridge hands a day, a perfect deal would occur only once in 124 trillion years!

238. Computers

Three computers.

239. A Third of the Planet

You would have to be at a distance equal to the Earth's diameter—about 7,900 miles.

Imagine Earth to be a circle instead of a sphere:

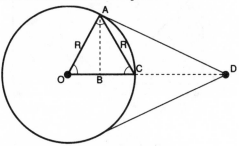

$\triangle$ OAC is equilateral.

$\triangle$ OAB is similar to $\triangle$ OAD. OD:R $= \dfrac{R}{2}$

$R^2 = $ OD: $\dfrac{R}{2}$ or OD $= 2R$

240. Checkers

Take the 2 checkers shown at left in the illustration, and shift them to positions at right, pushing their respective rows backwards to make the columns align.

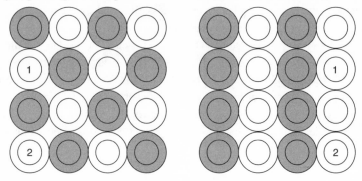

241. The Handicap Race

Mel wins again. In the second race, after Sid has gone 90 yards, Mel will have gone 100, and they will be alongside each other. There are 10 more yards to run, and since Mel is the faster runner, he will finish first.

242. . . . 9, 10

9 below, 10 above. Numbers appearing above the line are spelled with 3 letters only.

243. The South Pole

$-40°$ Centigrade $= -40°$ Fahrenheit.

244. Guinness or Stout

The second man put the 50 cents down in some combination of change—for instance, 4×10 cent and 2×5 cent coins—so that he could have put just 45 cents down if he had wanted Jubilee.

245. Bonus Payments

Let m be the number of men and let x be the fraction of men refusing a bonus. Then the amount paid out is given by

$$T = 8.15(350 - m) + 10(1 - x)m = 2852.50 + m(1.85 - 10x)$$

which will be independent of m only if $x = 0.185$, so that $T = 2852.50$. Both m and $0.185m$ are integers with $m \angle 350$, so $m = 200$. It follows that \$1,222.50 is paid to the 150 women.

246. The Tiled Floor

To fall and not touch a line the card must fall so that the center of the card falls within the shaded area.

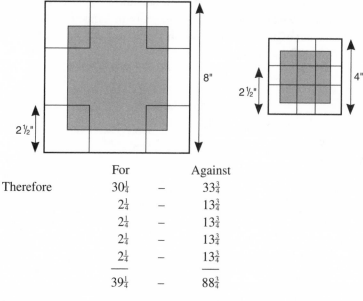

	For		Against
Therefore	$30\frac{1}{4}$	–	$33\frac{3}{4}$
	$2\frac{1}{4}$	–	$13\frac{3}{4}$
	$2\frac{1}{4}$	–	$13\frac{3}{4}$
	$2\frac{1}{4}$	–	$13\frac{3}{4}$
	$2\frac{1}{4}$	–	$13\frac{3}{4}$
	$39\frac{1}{4}$	–	$88\frac{3}{4}$

$\times 4$ to remove fractions = 157 to 355 against.

247. A Shuffled Deck

The number of red cards in the top 26 must always equal the number of black cards in the bottom 26. Hence, by the rules of logic, the statement is correct no matter what follows the word "then!"

248. A Peculiar Number

The common result must have 7 and 11 as factors, thus the number is $7 + 11$ or 18. The method is general, since the solution of $(x - k)k = (x - m)m$ is $k + m$.

249. Antifreeze

One quart of the old solution differs from one quart of the new (or average) solution by -24 percent, while one quart of the solution to be added differs from the new solution by $+48$ percent. Hence, there must be 2 quarts of the old solution for each quart of the added solution. So a third of the original radiator content or 7 quarts must be drained.

250. Tree Leaves

If it be required that no tree be completely devoid of leaves, then the statement would be true.

251. The Will

Daniel Greene's evident intention was that his estate be divided 2 : 1 between his son and his widow, or 1 : 3 between his daughter and his widow. These ratios can be preserved by giving the son six-tenths of the estate, the daughter one-tenth, and his widow, Sheila, three-tenths.

Let the value of the estate be S and   Sheila's share x

Boy's share y

Girl's share z

By definition Sheila is to receive half of the boy's share and three times the girl's share, or

$$x = \frac{y}{2}, \quad x = 3z$$

But $x + y + z = s$ or

$$x + 2x + \frac{x}{3} = s$$

$$\therefore \quad \frac{3x}{3} + \frac{6x}{3} + \frac{x}{3} = \frac{3s}{3}, \qquad \therefore \ 10x = s$$

The estate is therefore shared Mother $\dfrac{3}{10}$

Boy $\dfrac{6}{10}$

Girl $\dfrac{1}{10}$

252. Watered-down Wine

There is exactly as much water in the wine pitcher as there is wine in the water pitcher. Regardless of the proportions of wine and water which have been transferred, if both pitchers originally held equal volumes of unadulterated liquids and both are eventually left the equal volumes of mixtures, then equal amounts of wine and water must have been transferred.

This old brainteaser also forms the basis of a perplexing card trick: The performer and the spectator are seated opposite each other at a table. The performer turns 20 cards face-up from a pack of 52 cards. The spectator is asked to shuffle the pack so that the reversed cards are randomly distributed, then to hold the pack out of sight beneath the table and to count off 20 cards from the top. These 20 cards are passed, under the table, to the performer.

Having taken the 20 cards, the performer continues to hold them beneath the table, and tells the spectator: "Neither of us knows how many reversed cards there are in this pack of 20. However it is likely that there are fewer reversed cards in the pack of 20 than there are in the pack of 32 which you are holding. Without looking at my cards, I am going to turn some more face-down cards face-up in an attempt to equalize the number of reversed cards in my packet with the number in yours."

The performer then fiddles with his packet of cards under the table, making out that he can feel the difference between fronts and backs. After a few moments, he brings them into view and spreads them on the table. When the face-up cards are counted, it turns out that their number is exactly the same as the number of face-up cards in the spectator's packet of 32. What he's done, of course, is just turn his entire stack of 20 cards over!

253. A Logic Riddle

The answer is 4. The problem can be written by way of analogy as follows:

$$\frac{5}{2} : 3 = \frac{10}{3} : x$$

$$\therefore x = 4$$

An alternative reasoning goes as follows:

If $2\frac{1}{2} = 3$, then $10 = 12$

Therefore $\frac{1}{3}$ of 10 would be 4

254. A Matter of Health

If there were just two ailments with the percentages 70 percent and 75 percent, then the minimum overlap would be 45 percent, or 70 percent plus 75 percent minus 100. The minimum 45 percent of the population with the first two conditions similarly overlaps the 80 percent with the third ailment by a minimum of 25 percent. Finally the minimum of 25 percent suffering from the first three ailments overlaps the 85 percent with the fourth condition by at least 10 percent. The same principle would apply to any other combination of any number of ailments. The answer can be calculated instantly with the following formula:

$$(A1\% + A2\% + A3\% + \ldots An\%) - 100 \times (N - 1) = 10\%$$

where A1 . . . An are the percentages of the various ailments and N is the number of different ailments.

255. Streetcars

The solution normally given to this problem uses the method of relative speeds: the relative velocity between man and streetcar when going in the same and opposite directions respectively is proportional to the number of cars encountered. This establishes the equation:

$$(x + 3)(x - 3) = \tfrac{60}{40}$$

Therefore $x = 15$ miles per hour.

However, a simpler explanation that short-circuits the algebra is this: Picture two cars at the start of the walk, the 40th car behind the man and the 60th car ahead of him. These must obviously have each traveled half the distance between them when they met at the man, namely, a 50-car space. So the distance walked in the same period was a 10-car space, or one-fifth as much, which means the speed of the streetcar was 15 miles per hour.

256. Passing Trains

If it takes 10 seconds for a train of length L to pass A, and 9 seconds to pass B, the relative velocity between the train and A is L/10, between the train and B is L/9, and between A and B is L/90. The latter figure is one-tenth the relative velocity between the train and B. Since it took the train 1,210 seconds to reach B, it will take A ten times as long, or 12,100 seconds, of which 1,219 seconds had elapsed when the train passed B, leaving 3 hours, 1 minute and 21 seconds.

An alternative, and perhaps more interesting, solution considers an observer looking out of a supposedly stationary train at the two walkers. It appears to such an observer that the woman moves faster than the man, since the woman takes 9 seconds to cover a distance that the man covers in 10 seconds, and thus in a 9-second period the woman gains 1 second over the man. It is given that the woman goes past the rear end of the train 20 minutes and 9 seconds after the man, and for them to meet it would take 9 times this interval, or 3 hours, 1 minute and 21 seconds.

257. The Fly and the Record

It will arrive at the outer edge. When a record is played on a turntable, it revolves clockwise when seen from above, and relative to the record, the needle moves counterclockwise as seen from above. If the needle—and hence the fly—were to move clockwise around the groove, it would end up at the outer edge.

258. Move One Coin

Simply place one coin on top of another at the intersection of the two rows.

259. The Unbalanced Coin

An unbalanced coin can be used to generate a series of truly random numbers. In trying to determine each number, toss the coin twice. Since the coin is biased, the outcome heads-heads (HH) will not occur with the same frequency as tails-tails (TT). But the sequence HT is as likely as TH, no matter how unfair the coin may be. You simply flip the coin twice for each trial, rejecting both HH and TT, then designate HT as "one" and TH as "zero" (or vice versa).

260. Bicycle Experiment

Strange as it may seem, pulling back on the lower pedal causes the bicycle to move backwards. The force on the pedal is in the direction that would normally propel the bicycle forward, but the large size of the wheels and the small gear ratio between the pedal and the wheel sprockets are such that the bicycle is free to move backward with the pull. When it does so, the pedal actually moves forward with respect to the bicycle (that is, in a counterclockwise direction in the illustration), although it moves backward with respect to the ground.

The higher pedal, if pulled back, would simply free-turn until it reached the point in its arc closest to the source of the pulling-force, at which point the bike would move backwards.

261. A Piece of String

Assume that at A the rope going from top left to bottom right is on top. (If it is the other way a mirror solution is produced which does not alter the probability.)

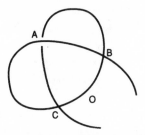

Then we have four equal possibilities. The segment BOC can be either on top or underneath at each of the points B and C.

Segment BOC

At B	*At C*	*The rope is*
On top	On top	Not knotted
On top	Under	Not knotted
Under	On top	Knotted
Under	Under	Not knotted

Hence the chance that the rope forms a knot is $\frac{1}{4}$.

262. The Island and the Trees

The girl ties one end of the rope to the tree at the edge of the lake. She walks around the lake holding the other end of the rope, then ties that end to the same tree. The doubled rope is now firmly stretched between the two trees, making it easy for her to pull herself through the water, by means of the rope, to the island.

263. A Boy, a Girl and a Dog

The dog can be at any point between the boy and the girl, facing either way. To prove this, at the end of one hour place the dog anywhere between the boy and the girl, facing in either direction. Reverse all the motions, and all three will return at the same instant to the starting point.

264. Boxes and Balls

Once a girl removed 2 balls from her box, she narrowed the possible combinations in her box to 2. If she was able to deduce the color of the third ball, it must have been because the label on her box showed one of the 2 possible combinations, forcing the actual contents to be the other possible combination—remember, all the boxes were incorrectly labeled.

The first girl's box contained either BBB or WBB, and its label must have read WBB or BBB for her to have guessed the color of the third ball. Similarly, the second girl's box must have contained either WBB or WBW, with its label reading either WBW or WBB.

The third girl's box must have contained either WWW or WBW—but her box could not have been labeled either of these, otherwise she, too, would have been able to deduce the color of her third ball.

The only distribution satisfying all these conditions is:

Girl	1	2	3	4
Label	W B B	W B W	B B B	
Actual	B B B	W B B	?	

The fourth girl instantly realized that her box must be labeled WWW; since it could not actually have contained WWW, the third girl's box must have contained that combination, leaving her (the fourth girl's) box to contain WBW.

265. Two Trains

The passenger train is 3 times as fast as the freight train. Using the formula:

$$\text{Time (T)} = \frac{\text{Distance (D)}}{\text{Speed (S)}}$$

Let X be the speed of the passenger train, and
Let Y be the speed of the freight train, and
Let T1 and T2 be the time taken for passing (overtaking and meeting respectively), then:

$$T1 = \frac{D}{X - Y} \text{ and } T2 = \frac{D}{X + Y}, \text{ also}$$

$T1 = 2 \times T2$, therefore:

$$\frac{D}{X - Y} = \frac{2.D}{X + Y} \quad \text{D cancels out, leaving:}$$

$$\frac{1}{X-Y} = \frac{2}{X+Y}$$ This simplifies to $X + Y = 2X - 2Y$, leaving:

$$X = 3Y$$

266. Missing Elevation

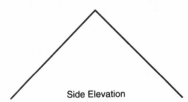

Side Elevation

This is a wire figure, in the form of an ellipse, bent at its smaller "diameter."

267. Avoiding the Train

The stretch of track the man was walking on was over a railway bridge or in a tunnel, and he was much nearer the end closer to the train than the farther end.

268. Bowl and Pan

She fills the pan on the table more than half full, and then carefully tilts up one end, pouring out the water, until the level reaches E, the bottom edge of the raised end (Figure 1). This leaves exactly half a pint in the pan, since the empty part is the same shape and size as the filled part.

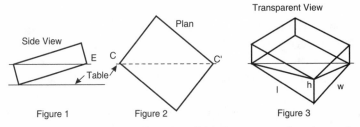

Side View

Plan

Transparent View

E C   C'

Table

Figure 1   Figure 2   Figure 3

The table has a straight edge, and mother slides the pan over the edge so that the opposite corners C and C′ coincide with the edge (Figure 2). She starts tilting again, with the bowl held to catch the water, and tilts until the surface of the water coincides with the corners C and C′. The bowl now contains one-third of a pint.

Proof: If the proportions of the pan are ignored, Figure 3 shows the pan at the end of pouring. The remaining water is in the form of a pyramid, the volume of which equals the area of the base times one-third its height.

$$\text{Area of base} = \frac{1w}{2}$$

$$\therefore \qquad \text{volume} = \frac{1w}{2} \times \frac{h}{3} = \frac{1wh}{6}$$

Since the volume of the pan (1 pint) = $1wh$, the remaining water = one-sixth of a pint. Thus, she poured into the bowl one-half less one-sixth of a pint, or one-third of a pint.

269. Jasmin's Age

Jasmin's age was 22 years and 8 months.

Let Jasmin's real age be "x."

She reduced her real age by $\frac{1}{4}$ minus 1 year.

$$\therefore \left(\frac{x}{4} - 1\right) + 18 = x \text{ or}$$

$$x - 4 + 72 = 4x$$

Reduce to

$$3x = 68 \qquad \text{or} \qquad x = 22\tfrac{2}{3} = 22 \text{ years and 8 months}$$

270. A Ball of Wire

This problem can be solved by reference to Archimedes' discovery that the volume of a sphere is two-thirds the volume of a cylindrical box into which the sphere exactly fits. The ball of wire has a diameter of 24 inches, so its volume is the same as that of a cylinder with height 16 inches and base diameter 24 inches.

Since wire is simply an extended cylinder, it is necessary to calculate how many pieces of wire 16 inches high and one-hundredth of an inch in diameter are equal in volume to a 16-inch-high cylinder with base diameter of 24 inches. Areas of circles are in the same proportion to each other as the squares of their diameters. The square of $\frac{1}{100}$ is $\frac{1}{10000}$, and the square of 24 is 576. Hence the cylinder is equal in volume to 5,760,000 of the 16-inch-long wires. The total length of the wire, therefore, is $5,760,000 \times 16$, or 92,160,000 inches = 1,454 miles and 2,880 feet.

271. Ferry Boats

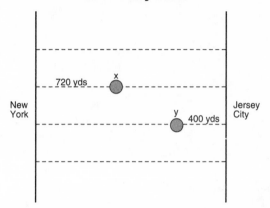

When the ferry boats meet at point X, they are 720 yards from one shore. The combined distance that both have traveled is equal to the width of the river. When they reach the opposite shore, the combined distance is equal to twice the width of the river. On the return trip, they meet at point Y after traveling a combined distance of 3 times the width of the river, so each boat has gone 3 times as far as they had when they first met.

At the first meeting, one boat had gone 720 yards, so when it reaches Z it must have gone three times that distance = 2,160 yards. This distance is 400 yards more than the river's width, which must therefore be $2,160 - 400 = 1,760$ yards or 1 mile wide.

272. John and the Chicken

First, determine how far John would travel to catch the chicken if the chicken and John both ran forward on a straight line. Add to this the distance that John would travel to catch the chicken if they ran towards each other on a straight line. Divide the result by 2 and you have the distance that John travels.

In this case, the chicken is 250 yards away, and the speeds of John and the chicken are in the proportion of 4 to 3. So, if both ran forward on a straight line, John would travel 1,000 yards to overtake the chicken. If they traveled towards each other, John would travel four-sevenths of 250, or $142\frac{6}{7}$ yards. Adding the two distances and dividing by 2 gives us $571\frac{3}{7}$ yards for the distance traveled by John. Since the chicken runs at three-fourths the speed of John, it will have traveled three-fourths of John's distance, or $428\frac{4}{7}$ yards.

273. The Prisoner's Choice

The prisoner believed he could improve his chances by distributing the gold and silver balls unevenly between the two urns. In fact, he decided his best strategy was to place just one silver ball in one of the urns and the remaining 49 silver balls and all 50 gold balls in the second urn. This way, if his random choice of urn brought him the second urn to pick from, his chances of picking a life-saving silver ball were only slightly worse than 1–2 (49–99), while if he was lucky enough to choose the first urn, he was certain to escape death.

In this manner, his overall probability of surviving was

$$\frac{1}{2} \times 1 + \frac{1}{2} \times \frac{49}{99} - \frac{74}{99}$$

or a little less than 3–4.

274. Counters in a Cup

There is no ordinary way in which the counters can be distributed to solve the problem, so there must be a catch. It lies in the ambiguity of how one thing can be placed "inside" another. The solution requires that one cup, containing an odd number of counters, be placed inside another cup (initially containing an even number of counters, which number is rendered "odd" by the counters in the inset cup). The illustration shows one of several ways in which the desired result can be achieved.

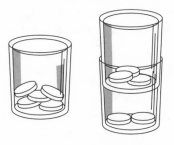

275. Speed of Ant

The ant is approaching Grand Central at 181 inches per second. It doesn't matter how fast the man eats. Since the ant is walking away from his mouth at 1 inch per second—i.e. 3 inches per second relative to the hot dog—it is moving towards the station 1 inch per second faster than the man is.

276. Wayne and Shirley

Most people think that two of each is more likely. But the correct answer is in fact three of one sex and one of the other. Set out below are all the possible combinations of four children:

B B B B

B B B G ⎫
B B G B ⎪   one
B G B B ⎬   girl
G B B B ⎭

G G G B ⎫
G G B G ⎪   one
G B G G ⎬   boy
B G G G ⎭

B B G G ⎫
B G B G ⎪
B G G B ⎬   two of
G B B G ⎪   each
G B G B ⎪
G G B B ⎭

G G G G

Each of the 16 arrangements is equally likely. In eight cases, there is a three-one split, whereas in only six cases is there a two-two split.

277. Word Series

The answer is (d) *heaven*. The sequence of ordinal numbers is implied: first aid, second nature, Third World, Fourth Estate, Fifth Column, sixth sense, seventh heaven.

278. Shooting Match

No. Bill and Ben's overall shooting performance was the same. Accuracy ratings are calculated from the ratio of hits to attempts. Bill's rating was 28/84 and Ben's was 25/75, so the two men tied because each hit the target with one third of his shots.

279. Lethargic Llamas

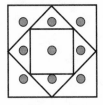

Two more square pens can be positioned as shown to give each llama its own enclosure.

280. Torpid Tapirs

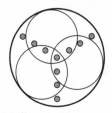

Three overlapping circular enclosures can be positioned as shown to separate the torpid tapirs.

281. Caroline and the Lake

If Caroline's objective had been to escape by reaching the shore as quickly as possible, her best strategy would have been to row so that the center of the lake, marked by the buoy, was always between her and her husband. At the same time, maintaining the three points in a straight line, she should have rowed towards the shore. Assuming that her husband followed the strategy of always running in the same direction around the lake, at a speed four times as fast as she can row, Caroline's optimum path was a semicircle with a radius of $\frac{r}{8}$, where r is the lake's radius. At the end of this semicircle, she would have reached a distance of $\frac{r}{4}$ from the center of the lake. That is the point at which the angular velocity she would have had to maintain to keep her husband opposite her just equals his angular velocity, leaving her no reserve energy for moving outward. (If during this period her husband changed direction, she could have done as well or better by mirror-reflecting her path.)

As soon as Caroline reached the end of the semicircle, she headed straight for the nearest spot on the shore. She had a distance of $\frac{r}{4}$ to go. Her husband would have had to travel a distance of π times r to catch her when she landed. She would have escaped, because when she reached the shore he would have gone a distance of only $3r$.

Suppose, however, Caroline preferred to reach the shore not as soon as possible but at a spot as far away as possible from her husband. In this case her best strategy, after she reached a point $\frac{r}{4}$ from the center of the lake, would have been to row in a straight line that is tangent to the circle of radius $\frac{r}{4}$, moving in a direction opposite to the way her husband was running.

282. Wooden Block

Diagonal grooves were cut in the wooden block as shown on the previous page. The two pieces could then slide apart diagonally.

283. Word Affinity
With the exception of SOCK, all the words in the two rows can be prefixed by AIR. (The prefix for SOCK that most immediately comes to mind is WIND.)

284. Gun Problem
I packed my gun diagonally in a flat square case with sides 1 yard long. The length of the diagonal was $\sqrt{2}$ yards or more than 1.4 yards.

285. Counter Colors
Most people think that because the state of the bag after the removal of the white counter is exactly the same as it was before the white counter was put in, the probability must be 1–2. This is not, however, the case.

The odds of the original counter (let's call it Counter x) in the bag being black or white are even. Adding a white counter (call this Counter y) makes the odds of the bag containing 2W or WB even. After one white counter is removed, three possibilities present themselves:

a) The counter removed was x, in which case the one left is certainly white.

b) The counter removed was y, and the one left is white.

c) The counter removed was y, and the one left is black.

In two out of the above three possible cases, the counter left in the bag is white; so the odds are 2 : 1 in favor of the second counter being white.

286. Bank Account
There is no reason whatever why the customer's original deposit of $100 should equal the total of the balances left after each withdrawal. The total of withdrawals in the left-hand column must always equal $100, but it is purely coincidence that the total of the right-hand column is close to $100. This is demonstrated by the following example, showing a different pattern of withdrawals:

Withdrawals	Balance left
$ 5	$ 95
$ 5	$ 90
$ 90	$ 0
$100	$185

287. Hat In the River

Because the rate of flow of the river has the same effect on both the boat and the hat, it can be ignored. Instead of the water moving and the shore remaining fixed, imagine the water as perfectly still and the shore moving. As far as the boat and the hat are concerned, this situation is exactly the same as before. Since the man rows 5 miles away from the hat, then 5 miles back, he has rowed a total distance of 10 miles with respect to the water. Since his rowing speed with respect to the water is 5 miles an hour, it must have taken him 2 hours to go the 10 miles. He would therefore recover his hat at 4 o'clock.

288. Tossing Pennies

No, it would be very unwise of Jack to accept the bet. To find the chances that the 3 coins will fall alike or not alike, consider all the possible ways that the 3 coins can fall, as follows:

1:	H	H	H
2:	H	H	T
3:	H	T	H
4:	H	T	T
5:	T	H	H
6:	T	H	T
7:	T	T	H
8:	T	T	T

Each of the 8 possibilities is equally likely to occur.

Note that only 2 of them show all the coins alike.

This means that the chances of all 3 coins being alike are 2 out of 8, or one quarter. There are 6 ways that the coins can fall without being all alike. Therefore the chances that this will happen are three-quarters.

In other words, Jill would expect in the long run to win 3 times out of every 4 tosses. For these wins Jack would pay her $1.50. For the one time that Jack would win, she would pay him $1. This gives Jill a profit of 50 cents for every 4 tosses on average.

289. The Kings

Let the 6 cards be numbered 1 to 6, and assume that the two Kings are cards 5 and 6. Now list all the different combinations of 2 cards that can be picked from 6, as follows:

1–2	2–3	3–4	4–5	5–6
1–3	2–4	3–5	4–6	
1–4	2–5	3–6		
1–5	2–6			
1–6				

Note that the Kings (cards 5 and 6) appear in 9 out of the 15 pairs. Since each pair is equally likely, this means that in the long run a King will be turned up in 9 out of every 15 tries. So the chances of getting a King are three-fifths. This of course is better than one-half, so the answer is that (a) is more likely.

290. Traffic Lights

The answer is $\frac{1}{8}, \frac{1}{4}, \frac{3}{8}, \frac{1}{4}, \frac{1}{8}$.

Since Robert traveled through the whole system in less than 2 minutes, the total distance is less than $2\frac{1}{2}$ miles, and no section is longer than $1\frac{3}{4}$ miles. If we chart the three arrival times at all possible positions of the first light (green from 3–16 seconds, 29–42 seconds, etc.), the only one allowing all three to pass is $\frac{1}{8}$ mile:

First light	Arrival time at 30 m.p.h.	Arrival time at 50 m.p.h.	Arrival time at 75 m.p.h.
$\frac{1}{8}$	15 seconds	9 seconds	6 seconds
$\frac{1}{4}$	30 seconds	18 seconds (RED)	
$\frac{3}{8}$	45 seconds (RED)		
$\frac{1}{2}$	60 seconds	36 seconds	24 seconds (RED)
$\frac{5}{8}$	75 seconds (RED)		
$\frac{3}{4}$	90 seconds	54 seconds (RED)	

Robert arrives at the last light just as it changes. A table of each $\frac{1}{8}$ mile together with the light sequence times of traffic lights (should they be situated there) shows that the only distance where a light change coincides with Robert's arrival is $1\frac{1}{4}$ miles after the start. (For example, a light at $\frac{1}{4}$ of a mile after the start would have green showing 15 seconds later than at the first light, and a light at $\frac{3}{8}$ of a mile would have green showing 30 seconds later, and so on.)

The same chart shows that, as Robert is not stopped, there is no light at $\frac{1}{4}$, $\frac{5}{8}$ or 1 mile from the start. The information about Hubert enables the rest of the distances to be calculated.

291. The Feast Day

Assume there are n days between consecutive Feast Days, that the temple bell rings every x minutes, and the monastery bell rings every $x + p$ minutes. Since the two bells alternate, the situation is as follows:

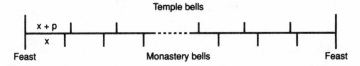

The pauses between successive rings are p, $2p$, $3p$, . . . , $3p$, $2p$, p. Since one of these is 1 minute, it follows that $p = 1$. Since the first temple bell is 1 minute after the monastery bell, the second temple bell is 2 minutes after the monastery bell, and the xth temple bell is x minutes after the xth monastery bell.

Therefore, in the $n \times 24 \times 60$ minutes between Feast Days, there are exactly x intervals of $x + 1$ minutes. Hence $x(x + 1) = 1440 \times n$.

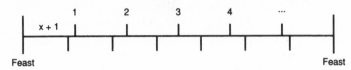

The problem is therefore to find a prime number n such that $1440 \times n$ is the product of two consecutive integers. The obvious candidates are $n = 1439$ and $n = 1441$, and indeed 1439 is prime. Hence the answer is 1,439 days.

292. The Clock-mender

The answer is 9:48 A.M. on the previous Monday.

We know from the question that the period of observation must be less than 8 days, and since the two clocks are known to be keeping different times and gaining or losing less than 60 minutes per day, they cannot both be gaining or both losing, for the faster clock could not overtake the slower clock by 12 hours in so short a period. Hence one clock must be gaining by, say, x minutes per day, and the other losing by, say, y minutes per day. Thus after a true elapsed time of m minutes, the two clocks will have moved forward respectively by:

$$m \times \frac{1440 + x}{1440} \text{ and } m \times \frac{1440 - y}{1440}$$

minutes. For both to show the same hour, the difference between these two movements must equal 12 hours or some multiple thereof, i.e.:

$$\frac{m(x + y)}{1440} - 720 \text{ etc. or } m - \frac{720 \times 1440}{(x + y)} \text{ etc.}$$

Since m is known to be an integer and less than 8×1440 minutes, $(x + y)$ must be a factor of 720×1440, which exceeds 90 but is less than 120, since x and y are each less than 60. The only such factors are 96, 100 and 108.

If $(x + y)$ were 96, the true elapsed time, m, would be 10,800 minutes, or 7 days plus 12 hours, which would have terminated outside the clock-mender's working hours so the coincidence of the clocks would have been unobserved. Similarly if $(x + y)$ were 108, the true elapsed time would be 9,600 minutes or 6 days plus 16 hours, which again would have terminated outside working hours, since 5:00 P.M. to 9:30 A.M. is already $16\frac{1}{2}$ hours. Thus $(x + y)$ can only be 100, giving a true elapsed time of 7.2×1440 minutes, or $7\frac{1}{5}$ days. Since both clocks have moved forward an exact number of minutes, both x and y must be multiples of 5.

The alternatives, therefore, for the clock which is gaining are 55, 50 or 45 minutes per day, corresponding to gains over the period of 6 hours 36 minutes, 6 hours, and 5 hours 24 minutes respectively, or total forward movements of (7 days plus) 11 hours 24 minutes, 10 hours 48 minutes and 10 hours 12 minutes respectively. Since the clocks were showing a time of 8 o'clock, these movements correspond with original setting times of 8.36, 9:12 and 9:48, of which only 9:48 the previous Monday morning lies within the clock-mender's working hours.

293. The Typewriter

The answer is 1 3 1 2 2 2 2 3 2 1.

The old keyboard layout was:

Row 1	Q W E R T Y U I O P
Row 2	A S D F G H J K L
Row 3	Z X C V B N M

The new keyboard layout is:

Row 1 K[d] C[e] L[f] A[f] G[f] V[g] F[k] N[k] X[l] J[l]

Row 2 B[a] E[a] R[a] P[h] I[h] W[dj] Y[dj] Z[l] M[l]

Row 3 S[b] T[b] O[b] U[b] H[c] D[i] Q[j]

The letters in square brackets indicate which of the following clues can be used to deduce the positions of the letters on the new keyboard.

(a) BEER (1 row)

Old positions: 3 1 1 1

So on the new keyboard, all must be in row 2.

(b) STOUT (1 row)

Old positions: 2 1 1 1 1

So on the new keyboard, all must be in row 3.

(c) SHERRY (2 rows)

Old positions: 2 2 1 1 1 1

New possibilities are 3 3 2 2 2 (2 or 3)

H could have been in row 1 or row 3, but since only 2 and 3 can be used, H must be in row 3.

(d) WHISKEY (3 rows)

Old positions: 1 2 1 2 2 1

New possibilities are (2 or 3) 3 (2 or 3) 3 1 (2 or 3).

K could have been in row 1 or row 3, but since all three rows must be used, K must be in row 1. At least one of W, I and Y must be in row 2, and any not in row 2 must be in row 3.

(e) HOCK (2 rows)

Old positions: 2 1 3 2

New positions are 3 3 1 1.

C could have been 1 or 3, but 3 is barred.

(f) LAGER (2 rows)

Old positions: 2 2 2 1 1

New positions are 1 1 1 2 2.

L, A and G could have been all in row 1 or all in row 3, but since there is no room in row 3, they are all in row 1.

(g) VODKA (2 rows)

Old positions: 3 1 2 2 2

New positions are 1 3 (1 or 3) 1 1.

V could have been in row 1 or row 2 but only rows 1 and 3 can be used.

(h) CAMPARI (2 rows)

Old positions: 3 2 3 1 2 1 1

New positions are 1 1 (1 or 2) 2 1 2 2.

P and I could have been in row 2 or row 3, but only rows 1 and 2 can be used.

(i) CIDER (3 rows)

Old positions: 3 1 2 1 1

New positions are 1 2 3 2 2.

D must be in row 3 as all three rows must be used.

(j) SQUASH (2 rows)

Old positions: 2 1 1 2 2 2

New positions are 3 3 3 1 3 3.

Q fills row 3.

(k) FLAGON (2 rows)

Old positions: 2 2 2 2 1 3

New positions are 1 1 1 1 3 1.

F could have been in row 1 or row 3, but 3 is full. N could have been in row 1 or row 2, but only two rows can be used.

(l) MUZZY (2 rows)

Old positions: 3 1 3 3 1

New positions are 2 3 2 2.

Thus the remaining letters, X and J, must be in row 1.

294. The Bridge

Michael set out for B-town as soon as he saw the sentry disappear into his bunker. Timing his progress, he walked for almost 5 minutes. He then turned round and started running back towards A-town. The sentry emerged and, seeing Michael running towards A-town, ordered him to "return" to B-town.

295. The Cookie Jar

While this is ostensibly a "trial-and-error" exercise, a systematic approach is possible. Let us assume, in turn, that each child has stolen the cookie, and see whether the other statements are then compatible with the condition that three are lies and only one statement is true.

If Ann is the thief, her statement is a lie; Harry's statement is a lie; Lisa's statement is true; Fred's statement is true.

Therefore this cannot be the solution.

In practice, it saves time to test the person who makes a statement about herself—in this case Lisa. If Lisa is the thief, then Ann's statement is a lie; Harry's statement is a lie; Lisa's statement is a lie; Fred's statement is true.

296. Crossing the Desert

Each man will have: 2 full bottles, 1 half-full bottle, and 2 empty bottles.

Reasoning: There is enough water for $7\frac{1}{2}$ full bottles. There are 15 bottles altogether. Therefore each man will end up with $2\frac{1}{2}$ full bottles and $2\frac{1}{2}$ empty bottles. However, half an empty bottle is the same as half a full bottle, leading to the above result.

297. Panama Canal

The west end of the Panama Canal is in fact in the Caribbean and the east end is in the Pacific. The confusion arises because the isthmus curves around at that point. As can be seen from any atlas, the canal runs from north-west to south-east.

298. The Short Cut

At 40 miles per hour, the train would enter the tunnel when John was still two-eighths from the exit or a quarter of the tunnel's length. If the train was to reach him at the exit, it would have to travel at four times John's speed, i.e. 40 miles per hour.

299. Red, White and Blue

There are six possible pairings of the two balls withdrawn:

RED   + RED

RED   + WHITE

WHITE + RED

RED   + BLUE

BLUE   + RED

WHITE + BLUE

We know that the WHITE + BLUE combination has not been drawn. This leaves five possible combinations remaining. Therefore the chances that the RED + RED pairing has been drawn are 1 in 5.

Many people cannot accept that the solution is not 1 in 3, and of course it would be, if the balls had been drawn out separately and the color of the first ball announced as red before the second had been drawn out. However, as both balls had been drawn together, and then the color of one of the balls announced, then the above solution, 1 in 5, must be the correct one.

300. Common Factor
Each word contains 3 consecutive letters from the alphabet.

301. Word Groups
The word DUNE from Group 2 belongs best with the words from Group 1, all of which may be preceded by the word "Sand."

302. Two Wins
If Bill is to win two games in a row, he must win the second game, so it is to his advantage to play that game against the weaker opponent. He must also win one game against the stronger opponent, and his chance is greater if he plays the stronger opponent twice. The first game should therefore be against his mother.

303. Find X
If each side is squared:

$$x + \sqrt{x + \sqrt{x + \sqrt{x}} \ldots} = 4$$

and if, as is stated:

$$\sqrt{x + \sqrt{x + \sqrt{x}} \ldots} = 2$$

then $x + 2 = 4$; so $x = 2$.

304. Pocketful of Coins
The lowest number of coins in a pocket is 0. The next greater number is at least 1, the next at least 2 and so on until the number in the ninth pocket is at least 9. Therefore the number of coins required is at least:

$$0 + 1 + 2 + 3 + 4 + 5 + 6 + 7 + 8 + 9 = 45$$

Since Freddy has only 44 coins, the answer is no.

305. Six-gallon Hat

The key is to reduce the content of container A to 6 gallons.

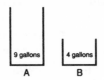

This can be achieved as follows:

1. Fill A.
2. Pour 4 gallons into B.
3. 5 gallons remain in A.
4. Empty B.
5. Refill B from A—this leaves 1 gallon in A.
6. Empty B and put the 1 gallon from A into B.
7. Refill A.
8. Fill B from A. This will take 3 gallons, leaving 6 in A.

306. Flock of Geese

We know each goose was sold for the same number of dollars as there were geese in the flock. If the number of geese is n, the total number of dollars received was n^2. This was paid in $10 bills plus an excess in coins. Since George drew both the first and last $10 bills, the number of $10 bills must have been odd, and since the square of any multiple of 10 contains an even number of tens, n must end in a digit, the square of which contains an odd number of tens. Only two digits, 4 and 6, have such squares: 16 and 36. Both squares end in 6, so n^2 is a number ending in 6. Thus the excess amount consisted of 6 dollars.

After Guy took the $6, he still had $4 less, so to even things up the older brother wrote out a check for $2.

307. Three Points on a Hemisphere

The probability is 1 (complete certainty). Any three points on a sphere must be on a hemisphere.

308. Deal a Bridge Hand

The dealer dealt the bottom card to himself, then continued dealing from the bottom counter-clockwise.

309. The Fifty-dollar Bill

Since the counterfeit note was used in every transaction, they are all invalid. Therefore, everybody stands in the same position to his/her creditor as before the banker picked up the counterfeit note.

310. The Bicycle Race

12 minutes.

311. The North Pole

If West and East were stationary points, and West on your left when advancing towards North, then, after passing the Pole and turning around, West would be on your right. But West and East are not fixed points, but *directions* round the globe. So wherever you stand facing North, you will have the West direction on your left and East on your right.

312. Card Games

Nine. Jack wins 3 games and thus gains $3. Jill has to win back this $3, which takes another 3 games, then win a further 3 games.

313. Long-playing Record

About 3 inches. The needle moves from the outermost position to the center area of the label in an arc whose radius is the length of the pick-up arm.

314. Which Games?

One friend plays none of the games, so the other two must each play all three.

315. What Day Is It?

Today is Sunday.

The drawing below will assist:

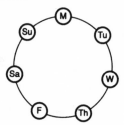

Choose a day at random, say Tuesday. Then:

1. When the day after tomorrow (Thursday) is yesterday (Monday), 4 days will have elapsed.
2. When the day before yesterday (Sunday) is tomorrow (Wednesday), 3 days will have elapsed.
3. Today will therefore be 7 days from Sunday; i.e. Sunday.

316. Cash Bags

The 10 bags should contain the following amounts: $1, $2, $4, $8, $16, $32, $64, $128, $256, $489.

317. Garage Space

Let the number of trucks be x. Then the garage had room for $(x - 8)$ trucks. By increasing the garage by 50 percent, there is now room for $(x + 8)$ trucks.

$$(x-8)+\frac{(x-8)}{2} = x+8$$
$$3(x-8) = 2(x+8)$$
$$3x-24 = 2x+16$$
$$x = 40$$

318. Bag of Chocolates

	Joan	Jane	Jill
	4 chocolates	3 chocolates	$\frac{14}{3}$ chocolates
or	12	9	14
Total	264	198	308
Ages	6	$4\frac{1}{2}$	7

319. Lost

Stand the signpost back up so that the arm indicating the place you just came from points in that direction. All the other arms will then point in the correct direction.

320. Problem Age

The statement was made on 1 January. Peter's birthday is 31 December. He was 17 the day before yesterday. Yesterday, the last day of last year, was his 18th birthday. He will be 19 on the last day of this year and 20 on the last day of next year.

321. Mate in One

Since chess is played with a white square at each player's near right corner of the board, the players must be at the left and right sides of the board illustrated. Therefore, whether White is moving to the left or to the right, he wins by queening his Pawn: P-B8 (Queen), mate.

322. Endgame

The key to this endgame lies in the fact that a player can elect what piece to promote an advanced Pawn to—usually a Pawn is promoted to Queen, but it does not have to be.

White advances his Pawn to B7. If Black does the same, to K7, then White will queen first and mate at R5.

If Black wants to prevent queening of the white Pawn, he might try B-B1. The Bishop is then taken by the white King, and White then proceeds to P-Q8, but promotes to a Knight rather than a Queen. When Black queens his pawn, White's Knight checks at K6 and mates with Bishop at K7.

323. Smart Kid

Let us call the experts Mr. White and Mr. Black, according to the color of the pieces each played against my daughter. Mr. White played first. My daughter copied his first move as her opening against Mr. Black at the other board. When Mr. Black had answered this move, she copied his move at the first board as her reply to Mr. White. And so on. In this way, the simultaneous games against the two experts became a single game between them; my daughter served as a messenger to transmit the moves. Hence she was certain that she would either win one game and lose the other, or draw both.

324. On the Move

    wP moves to 4;
    bP jumps to 3;
    bK moves to 5;
    wP jumps to 6;
    wK jumps to 4;
    wR moves to 2
    (the crucial move);
    bP jumps to 1;
    bK jumps to 3;
    bR jumps to 5;
    wP moves to 7;
    wK jumps to 6;
    wR jumps to 4;
    bK moves to 2;
    bR jumps to 3;
    wR moves to 5.

325. Checkmate

Remove the white Pawn from B6 to K4 and place a black Pawn on Black's KB2. Now White plays P to K5, check, and Black must play P to B4. Then White plays P, takes P en passant, checkmate. This was therefore White's last move, and leaves the position shown. It is the only possible solution.

326. White to Move

The key to this problem lies in considering the moves that could have led to the position shown. Black could not have just moved his King. If the King had moved from any other square, it would already have been in check prior to White's previous move. He could not have moved his Bishop or the three Pawns on their starting squares. Black's Pawn on B4 could not have moved from B3 because White would have been in check, and could not have arrived by capturing a piece. The only possible move Black could have made is P(B2)–B4. Therefore, White wins by P × P (en passant), mate.

327. Charles XII

You are on your own.

328. Railways

Rails expand with rising temperature and, traditionally, gaps between rail sections have been considered an essential element of track design to accommodate this expansion and prevent buckling.

If the temperature of x feet of rails is raised by t degrees, then the length increases by an amount of $0.000006\,xt$.

However, since the 1950s, this conventional wisdom has undergone some rethinking. Nowadays, it is accepted that rails can be welded together into lengths up to half a mile and even longer, without a break, increasing stability by anchoring the rails more closely to the ties; and laying rails at a time when temperature is close to, or slightly in excess of, the mean temperature of the site, it has proven possible to reduce the effects of heat expansion significantly.

329. Rice and Salt

Rice is more hygroscopic than salt and therefore tends to keep the salt dry.

330. Coal and Lime

The lime solution covers the coal with a white film. If coal is stolen after loading and spraying, it is evident that the theft was committed en route and is easily discovered as the train passes stations on the way to its destination.

331. Sand on the Beach

Before you step on it, the sand is packed as tightly as it can be under natural conditions. Your weight disturbs the sand, making the grains less efficiently packed. The sand is forced to occupy more volume, and rises above the water level, becoming dry and white. The water rises more slowly, by capillary action, so it takes a few seconds or more before the sand gets wet and dark again.

332. Bridge Clearance

The child's solution was elementary. "Let some air out of your tires," she said, "until the truck is low enough to pass under the bridge."

333. A Flat Tire

The man took one lug nut from each of the other three wheels, and attached the spare wheel. Each wheel was then held on by 3 nuts rather than 4, which was sufficient to keep the wheels on until the man reached the next town.

334. The Biology Exam

The student said, "Good, I didn't think so." He then jammed his test booklet into the middle of the stack and dashed out of the room. At least, so, proverbially, the story goes.

However, I can think of a sequel. The professor, a member of MENSA, refused to be outwitted by the smart-ass of a student. After marking the papers, he invited the students to collect their grades in person.

335. The Manhole

If, while being maneuvered, a square manhole cover happened to be turned on its edge, so that it was presented diagonally to the manhole, it would fall through the manhole.

336. Wet Roof

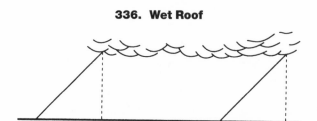

The figure above shows the rain from a 10-mile stretch of clouds falling at an angle of 45°. It wets the same extent of ground as it would if it fell vertically, but moved to the left. The point here is that the ground area is being compared with the cloud area, to which it is parallel, but in the case of the tilted roof described in the question, it was not. Thus, 10 miles of cloud wet 10 miles of ground, whereas 10 inches of cloud wet 14 inches of roof.

337. Interior Angles

Given △ ABC in the figure below, draw CE parallel to side AB, and extend BC to D. It is evident that ∠ 1 = ∠ 4, since they are alternate interior angles, and also that ∠ 2 = ∠ 5, since it can be seen that the sum of ∠s 1, 2 and 3 equals the sum of ∠s 4, 5 and 3, which, in turn, must equal 180° (as the latter three angles form a straight line).

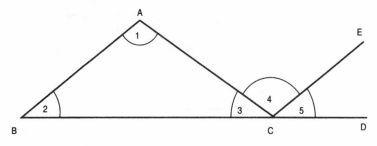

338. Angle in Semicircle

Let a semicircle be drawn with center O and diameter BC, and choose any point A on the semicircle. We must prove that $\angle$ BAC is a right angle.

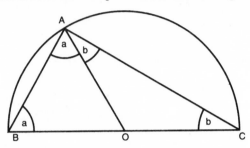

Draw line OA and consider $\triangle$ AOB. Since OB and OA are radii of the semicircle, they have the same length, and so $\triangle$ AOB is isosceles. Hence, $\angle$ ABO and $\angle$ BAO are equal; call them both a. Likewise, in $\triangle$ AOC, OA and OC have the same length, and so $\angle$ OAC = $\angle$ OCA; call them both b. Since the angles of a triangle always add up to 180°, we know that, in the large $\triangle$ ABC:

$$a + b + a + b = 180°$$
$$\therefore \qquad a + b = 90°$$

And $\qquad a + b$ is the same as $\angle$ BAC

339. Pythagoras

There are no fewer than 366 ways of proving the Pythagorean theorem (see Elisha S. Loomis, *The Pythagorean Proposition*). Probably the simplest is given in *Mathematical Quickies* by Charles W. Trigg:

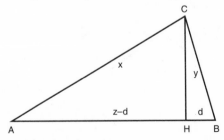

In the right $\triangle$ ACB, draw CH, the altitude to the hypotenuse AB. Call AB = z. This creates three similar triangles:

ACB, AHC and CHB (i.e. triangles whose angles match), from which the following can be deduced:

$$d:y = y:z, \text{ therefore } y^2 = dz \ldots (1)$$
$$\text{and } (z - d):x = x:z \text{ or } x^2 = z^2 - dz \ldots (2)$$
$$\text{add (1) and (2) therefore } y^2 + x^2 = z^2$$

340. Lunes

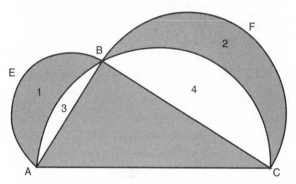

If length of $AC = x$
$AB = y$
$BC = z$

As area of two circles or semicircles are to each other as the squares of their diameters.

$$\widehat{AEB} : \widehat{ABC} = y^2 : x^2 \quad \text{and}$$
$$\widehat{BFC} : \widehat{ABC} = z^2 : x^2$$

Then: $\quad x^2(\widehat{AEB}) = y^2(\widehat{ABC}) \ldots (1) \quad \text{and}$

$$z^2(\widehat{ABC}) = x^2(\widehat{BFC}) \quad \text{or} \quad x^2(\widehat{BFC}) = z^2(\widehat{ABC}) \ldots (2)$$

Add (1) and (2):

$$x^2(\widehat{AEB} + \widehat{BFC}) = \widehat{ABC}(y^2 + z^2)$$

But $y^2 + z^2 = x^2$ (Pythagoras); thus $\widehat{AEB} + \widehat{BFC} = \widehat{ABC}$

Deducting areas 3 and 4 from each side results in:

Areas of Lune 1 + Lune 2 = $\triangle$ ABC

341. Diagonals of a Rectangle

A circle is the locus of all points which are equidistant from a given point (the center) O. Consequently points A, B, C, D are located on a circle.

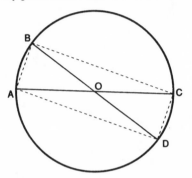

As all angles at A, B, C and D are inscribed in a semicircle, they are right angles.

342. Division of Angle Bisector

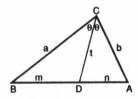

The areas of triangles with equal altitudes are proportional to their bases. Triangles BCD and ACD with bases m and n, respectively, have the same altitude. All points on the bisector of an angle are equidistant from the sides of the angle, so altitudes from D to BC and to AC are equal. Consequently:

$$m/n = BCD/ACD = a/b$$

343. Inscribed Decagons

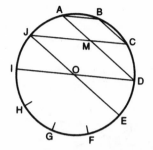

Diameters ID and JE are respectively parallel to the sides AB and BC of the regular decagon and to sides JC and AD of the star decagon. Hence ABCM and JMDO are rhombuses, so AD − BC = AD − AM = MD = JO, a radius of the circle.

344. Three Circles

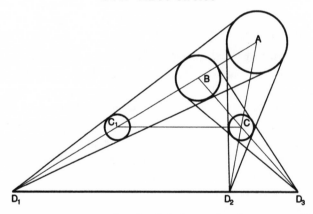

Draw common tangents to each pair of circles so that they meet at D_1, D_2 and D_3. Between the common tangents of A and B,—draw circle C_1 equal in size to C. Then:

$$\text{Radius A/Radius C} = AD_1/C_1D_1 = AD_2/CD_2 \text{ or}$$

$$AD_1/AD_2 = C_1D_1/CD_2$$

Therefore C_1C must be parallel to D_1D_2, and similarly C_1C must be parallel to D_1D_3. Therefore D_1D_2 and D_1D_3 are coincident and are one straight line.

345. A Thousand Points

It can be proven that for any finite set of points on a plane there is an infinity of straight lines that divide the set exactly in half. The following proof for six points can be applied to any finite number of points:

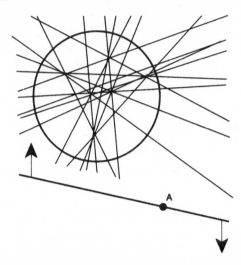

Construct lines through every pair of points. Select a new point, A, lying outside the closed curve but *not* lying on any of the lines. Draw a line through point A. If this line is rotated about point A, clockwise as shown, it passes over one point at a time. (It cannot pass two points simultaneously because that would mean that point A lay on one of the lines determined by those two points.) After the line has passed over half the points inside the curve, it will divide the set in half. Since A can have an infinity of positions, there must be an infinity of such lines.

346. The Hermit

Imagine two hermits walking on the same day, one up and one down, both following precisely the path that the real hermit had taken and both proceeding at the same rate of speed as the real hermit. The two imaginary hermits must meet somewhere along the path (though we can't say precisely where), and that is the spot the hermit had occupied on both trips at exactly the same time of day.

347. Volume of a Sphere

Imagine the sphere to be completely filled with tightly packed cones. Then the volume of the sphere can be approximated by:

Volume of cone 1 + cone 2 + cone 3 . . . cone n. As the bases of the cones become progressively smaller the heights approach the sphere's radius and the sum total of the bases of the cones approach the surface of the sphere, thus: $R^2\pi$ (the surface) times the height of the cones, R, divided by 3 becomes the volume of the sphere.

$$\frac{4}{3} \cdot R^3 \pi \, .$$

348. Five Points in a Square

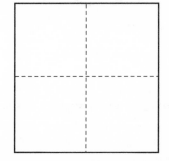

Draw two lines through the center of the square as shown. These two lines divide the square into four half-inch squares. At least two of the five points must be in or on the perimeter of one of the small squares. The maximum distance they can be apart is at the opposite ends of a small diagonal which would be $\frac{\sqrt{2}}{2}$ inches.

349. The Alternative Triangle

According to Pythagoras the triangle has a height of 12 inches. You can therefore construct a triangle:

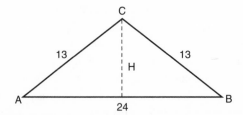

with a height of 5 inches and a base of 24 inches, each triangle having an area
of 60 square inches.

350. The Four Integers

Let X be any integer and the three consecutive integers be $(X + 1)$, $(X + 2)$, and
$(X + 3)$ respectively.

Then the product can be expressed as:

$$X (X + 1)(X + 2)(X + 3) = (X^2 + 3X)(X^2 + 3X + 2) = (X^2 + 3X + 1)^2 - 1$$

This cannot be a perfect square as two positive squares cannot differ by one.

351. Three Points on a Sphere

The key to the proof is the axiom that any three points must lie on a plane. Visualize a plane through the three random points and you will realize that they
must be on the same hemisphere.

352. The Three Cities

Taking the earth's circumference as 25,000 miles, cities A and B are anywhere
from zero to 7,000 miles apart, as the greatest distance between the three cities
must lie on a great circle, approximately equal to the circumference measured
at the equator.

353. Convergence

Let the limit of convergence be X, then

$$X^2 = 2 + \sqrt{2 + \sqrt{2 + \sqrt{2}}} + \dots therefore$$

:

$$X^2 = 2 + X \text{ or } X^2 - X - 2 = 0$$

This equation is satisfied by X = –1 or X = 2.

Rejecting the negative root we have proved that $\sqrt{2 + \sqrt{2 + \sqrt{2}}} + \dots$ converges to 2.

354. Circle and Point

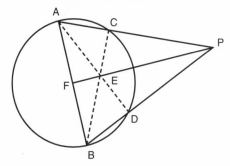

Draw lines from P to A and B intersecting the circle at C and D respectively. Then draw lines from A to D and B to C, intersecting at E. Now connect E with P which produces a line perpendicular to AB.

This proof is based on two theorems:

(a) Angles inscribed in semicircles are right angles.
(b) The altitudes of triangles meet in one point.

As PF is the altitude of triangles ABP the angle at F must be a right angle.

355. Area of Triangle

Convert the triangle into a parallelogram having the same base and height as the triangle.

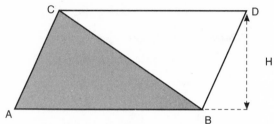

ABC is the triangle. ABDC is the parallelogram which has an area of the base AB times H and twice the area of the triangle.

Bibliography

General and Puzzles

Ball, W. W. R., and H. S. M. Coxeter, *Mathematical Recreations and Essays,* 12th edn, University of Toronto Press, 1974.

Barr, Stephen, *Miscellany of Puzzles: Mathematical and Otherwise,* Thomas Y. Crowell Co., 1965.

Brandes, Louis Grant, *Math Can Be Fun,* J. Weston Walch, 1975.

Dudeney, Henry E., *300 Best Word Puzzles,* Charles Scribner's Sons, 1972.

Emmet, E. R., *Puzzles for Pleasure,* Emerson Books, 1972.

Fixx, James F., *Games for the Superintelligent,* Doubleday & Co., 1972.

———. *More Games for the Superintelligent,* Doubleday & Co., 1975.

Gardner, Martin, *Mathematical Carnival,* Alfred Knopf, 1975.

———. *Mathematical Magic Show,* Alfred Knopf, 1977.

Graflund, Robert S., *Basic Math Puzzles: 50 Spirit Masters,* J. Weston Walch, 1976.

Herstein, I. N., and I. Kaplansky, *Matters Mathematical,* Harper & Row, 1974.

Holt, Michael, *Mathematical Puzzles and Pastimes,* Walker & Co., 1977.

Hunter, J. A. H., *Hunter's Mathematical Brain-Teasers,* Dover, 1977.

Hunter, J. A. H., and Joseph S. Madachy, *Mathematical Diversions,* Dover, 1975.

Kordemsky, Boris A., *The Moscow Puzzles,* Charles Scribner's Sons, 1972.

Madachy, Joseph S., *Mathematics on Vacation,* Charles Scribner's Sons, 1973.

Morris, Ivan, *The Pillow-Book Puzzles,* Bodley Head, 1969.

Rice, Trevor, *Mathematical Games and Puzzles,* Batsford/St. Martin's Press, 1974.

Scott, Joseph, and Lenore Scott, *Puzzles for Everyone,* Ace Books, Charter Communications, 1973.

Wickelgren, Wayne A., *How to Solve Problems: Elements of a Theory of Problems and Problem-Solving,* W. H. Freeman & Co., 1974.

Wyler, Rose, *Professor Egghead's Best Riddles,* Simon & Schuster, 1973.

Index of Puzzles

About the Author

Erwin Brecher was born in Budapest and studied mathematics, physics, engineering, and psychology in Vienna, Brno (Czechoslovakia), and London.

He joined the Czech army in 1938 and, after the Nazi occupation of Sudetenland, he escaped to England. Engaged in aircraft design during the war, Brecher later entered the banking profession, from which he retired in 1984. Currently he devotes his time to playing bridge and chess, and to writing books on scientific subjects or puzzles, such as his recent *Lateral Logic Puzzles, Surprising Science Puzzles,* and *The IQ Booster,* encompassing several of his interests.

Brecher published his first book in German during September 1995 and received the "Order of Merit" in gold from the city of Vienna in recognition of his literary achievements.

A member of Mensa, Erwin Brecher and his wife, Ellen, make their home in London, England.